WIND ENERGY IN AMERICA

WIND ENERGY IN AMERICA
A HISTORY

By ROBERT W. RIGHTER

UNIVERSITY OF OKLAHOMA PRESS : NORMAN AND LONDON

Text design by Bill Cason

Library of Congress Cataloging-in-Publication Data

Righter, Robert W.
Wind energy in America : a history / by Robert W. Righter.
p. cm.
Includes bibliographical references and index.
ISBN 0-8061-2812-7
1. Wind power—United States—History. 2. Electric power production—United States—History. I. Title.
TK1541.R54 1996
621.31'2136—dc20 95-24686
CIP

The paper in this book meets the guidelines for permanence and durability of the Committee on Production Guidelines for Book Longevity of the Council on Library Resources, Inc. ∞

1 2 3 4 5 6 7 8 9 10

For Sherry

CONTENTS

ILLUSTRATIONS

PREFACE

FROM AN ENVIRONMENTAL PERSPECTIVE, HUMAN SOCIETIES OF FUTURE centuries may recall the twentieth century with awe and some chagrin. Americans with the help of other Western, industrial nations, have consumed the earth's nonrenewable energy resources with remarkable alacrity. We have partaken of the feast, leaving our progeny with mere scraps. The energy banquet cannot go on forever. In the future, Americans will rely more on renewable energy sources— the kind of resources humans have relied upon for most all of history.

It is with these environmental concerns in mind that I offer this history of wind energy. Wind is a resource often in the human conscious, but seldom utilized in a constructive fashion. We often consider the wind as an annoyance, at times a danger, and occasionally a destroyer. It can demolish an umbrella, turn a gentle snow into a life-threatening storm, or whirl into a hurricane of unparalleled power. But along with these negatives there are positives: the wind can play to our aesthetic instincts. Sailing, for example, is more than a method of getting from one point to another. However, central to the point of this volume, the wind can provide human societies with needed power. Past generations understood these beneficial aspects of the wind better than do we. For centuries, wind powered globe-girdling sailing ships. Wind also played a significant role in the pumping of water, the grinding of grain and the performance of stationary rotary work. It even generated electricity. It is the initial purpose of this book to provide a narrative history of the American effort to utilize the wind for electrical energy.

Beyond that effort, this work seeks to demonstrate that electrical energy need not be produced solely from fossil fuels and other environmentally destructive sources. Environmentalists, worrying about the end of civilization, have perhaps co-opted for themselves the title of "the dismal science," from the economists. Such a doomsday might still come, but it could be delayed by sensible political policies and considerate technology.

In the telling of this story, a few themes emerge. As a non-scientist, I have come to realize that there is no perfect way to provide the energy that human beings want and need. No benign means to produce electrical energy exists. Each method takes its toll of either human or environmental resources, reminding one of the environmental cliché that for every gain there is a compensatory loss. But when the tally is complete, wind energy surely has merits that, relative to other sources of power generations, tip the balance in its favor. It is diffuse and difficult to harness, yet inexhaustible—and relatively benign. Our forefathers, who lived in a world less beset by scarcities (and scariness), did not value these merits. In fact, to them wind energy seemed dreadfully lacking in its contribution to the energy needs of a dynamic nation. The drawbacks of wind energy are still evident, but the gentle nature of this energy source has attractions that offer an appealing option to "the hard energy path."

Another theme that runs through the narrative is that of the friction between centralism and localism—an idea readily understood in politics that also resonates in the world of energy. From twelfth-century England to the present, individuals and corporations sought to monopolize and to centralize energy production. Scientists and industrialists dreamed of a huge, efficient energy system, and together, over the last one hundred years, they accomplished this goal. The nation is crisscrossed with networks of centrally owned, centrally controlled power. Wind energy, however, lends itself to individualism and decentralization, and as such has always represented an option to the primary thrust of U.S. technology. Nevertheless, it has been a negligible alternative, and in that respect the story of wind energy is not one of triumph but rather one of frustration; not one of consequence but rather one of antiquarian curiosity.

The last twenty-five years have profoundly changed the way Americans produce and consume electricity. Environmental concerns have altered profligate consumption patterns, and these same concerns have caused us to reconsider our dependence upon fossil fuel and nuclear energy. In a limited way, renewable sources have been introduced into the energy mix. Wind energy has come of age, providing significant kilowatt-hours of energy for human use. Practical application has finally intercepted technological theory. In this light, the story that unfolds in the following chapters is a contemporary one. One cannot ignore the historical background, but it must be noted that the marriage of wind and electricity is a relatively recent one.

My interest in wind energy first came to my attention in 1980 when I stayed for a couple of days with Bob Brown, a rancher in Hershey, Nebraska. Standing, rather forlornly I thought, in his ranch yard was an old steel structure. I took it to be a water-pumping windmill—traditional on the Great Plains—but Brown set me straight. He explained that it once held a wind turbine and that for some years it had supplied his family—himself, his wife, and children—with electricity. They had relied upon the windcharger, as ranchers often called them. Then after World War II, Rural Electrification Administration (REA) cooperative lines reached Brown's place. After some hesitation, the family hooked up to the "highline."

Even in the 1980s, Brown doubted that he had made the correct decision. Across the rolling hills from his ranch stood two tall smokestacks, emitting residues from a coal-fired generating plant. The electricity they provided him was becoming increasingly expensive.

Returning to my home in Wyoming, I pondered on Brown's situation. He had exchanged a free (or so it seemed), nonpolluting, renewable energy source for an increasingly expensive, polluting, nonrenewable source. Technological change always necessitates trade-offs, but I had believed the REA brought only benefits. I had not before heard of windchargers, an intermediate technology—such as Brown's—that had all but disappeared with the spread of centrally produced power. A little research revealed that ranchers used windchargers widely. However, information about these machines proved scarce and a narrative history did not exist, hence my ensuing research and this book.

The contemporary nature of the subject leads us to ask that difficult question: When does history end and current events begin? On this issue I must apologize to my historian colleagues. The later chapters of this work are not history. The evidence necessary to make reasonable judgments is not yet available. The final two chapters rely on wind-energy experts whose work and opinions I respect. The best one can do regarding the present is make informed guesses, and guessing is not one of the tenets of our profession.

I do believe, however, that a rationale exists in bringing this story to the present. The year 1960 might have provided a prudent breaking point between history and semispeculation—but to slight recent events would be to forfeit substantial readership. This was confirmed for me recently in an interview I conducted: I mistakenly stated that my work spanned a century, from 1880 to 1980, and the interviewee responded that if that was the case "I have nothing to say to you, since nothing happened before 1980." Of course, something did happen, but it was a prelude. To stop short of the last thirty years and to refuse to attempt making a reasonable construction of events would be—as my interviewee remonstrated—tantamount to playing the prelude without the opera.

Whatever the future fate of energy in the United States, there is little question that change will occur. Production of electrical energy will take a very different form in A.D. 2100, just as today's differs dramatically from that in 1900. How will human societies produce energy in 2100? It is impossible to say, but the world is already running out of nonrenewable sources. Even if geologists unearth vast new reserves, continued global reliance on fossil fuels and uranium may take us on a collision course with our own survival. It is reasonable to believe that renewable energy sources will become notable future players in electric energy production. One of those sources will be the wind, a resource long, long noted, but so little exploited.

ACKNOWLEDGMENTS

THE IDEA FOR THIS BOOK WAS BORN SOME YEARS AGO WHEN—AS MENTIONED in the Preface—I talked with Bob Brown, the Nebraska rancher who at one time had used a so-called windcharger to provide his electricity. Thinking it would be useful to write an article on electricity-generating windmills in the American West, I undertook some research. As often happens, the project grew, and with it my debt to many people, libraries, archives, and institutions. First I want to thank Bob Brown, and with him John Hill, professor emeritus of engineering at the University of Wyoming, for planting the idea.

Since I was neither aware of the subject nor familiar with the field of technology, my indebtedness has been magnified. The manuscript still suffers from my unversed and marginal understanding of technological things. However, whatever is right may be credited to such people as T. Lindsay Baker of Baylor University. From the moment I announced my intent, he has been a fount of historical information, sharing his knowledge unselfishly. C. L. Sonnichsen wrote of two types of historians: one type hoards information, the other shares everything with members of the history "club," fully appreciative that historiography is a shared undertaking, representing a communal effort. Baker certainly fits the latter mold. I also owe a particular debt to two professionals who truly understand wind turbines: Paul Gipe saved me from many errors, and as any reader of this book will see, I have relied greatly on his expertise. I thank him for being generous with his time, and patient with a true neophyte in a field where he is an acknowledged expert. My debt to Andrew Swift of the University of Texas at El Paso mechanical engineering department is also great. He patiently read the entire manuscript and saved me from a number of egregious technical errors. He was generous with his time, and often included me—a lone humanist—in activities and meetings with professionals relating to wind energy.

My wife, Sherry L. Smith, who also read the manuscript, claims no technological expertise. She has, however, offered much sympathetic criticism, saving me from embarrassing cliches and lapses in syntax and clarity. It is with pleasure and love that I dedicate this work to her.

Historians Charles Martin, Cheryl Martin, Wayne Fuller, Ron Weber, Chuck Ambler, and David Hackett—all colleagues at the University of Texas at El Paso—read chapters of the manuscript or listened with feigned interest about a subject from which their own historical interests were far removed. I am sure the book contains errors both in facts and ideas, in spite of the mobilization of all this expertise: these are mine alone.

I also want to thank numerous individuals, mainly Californians, who have offered their help or shared their opinions about wind energy. I refer to the many anonymous individuals who had no particular expertise on the subject but who, when asked, shared their opinions—some negative, some positive, but seldom neutral—about this relatively new phenomenon on the California landscape. They helped to construct the mosaic of opinion that is central to this study.

Of course, many individuals provided more than brief opinions regarding wind energy. It is their profession and their livelihood. I wish I had interviewed more people, but of those I did get to talk with I want to thank the following for giving of their time and sharing their insights: My long time friend Asa Barnes Jr.; Curt Maloy, president of the Desert Wind Energy Association; Bob Scheffler, Southern California Edison Company; Don Smith, consultant for Pacific Gas and Electric Company; Dave Kelly, field manager for SeaWest, Tehachapi; and David Spera, formerly of the Lewis Research Center, NASA. Paul Jacobs, the son of Marcellus Jacobs, Mrs. Marcellus Jacobs, and Fred Bruns, a Jacobs wind-turbine installer, all shared their knowledge of the formative years of the industry, and of Marcellus and the development of the renowned Jacobs wind turbine. Two philosophers, Ty Cashman and Gordon Brittan, provided particular insight, for they have combined an interest in the technology of energy with their academic training. Cashman clarified some of the California story; Brittan, a philosophy professor at Montana State University, gave invaluable assistance in understanding the saga of wind energy at Livingston, Montana. The late Dennis G. Shepherd, of Cornell University, shared his ideas and a manuscript with me, and talks with West Texans R. Nolan Clark, Vaugh Nelson, and Joe Spinhirne set me on the right path some years ago.

Others I wish to thank include Matthias Heymann, of the Deutsches Museum, Germany, Dave Kelly, field manager for SeaWest, Techachapi, and Curt Maloy, of the Desert Wind Energy Association, the trade organization at Palm Springs. Rural residents were always helpful in offering both remembrances and opinions. I particularly recall the late Ida Chambers, of Jackson Hole, Wyoming, a delightful lady. Her daughter-in-law, Becky Chambers, was equally gracious with time and information.

Without the financial support of universities it would be impossible for me to travel, research and ultimately write. The University of Wyoming, where I was formerly on the faculty, was instrumental in facilitating the initial effort, giving me two travel grants. The Wyoming Council for the Humanities provided generous assistance toward the research of an essay entitled "The Wind at Work in Wyoming." The American Historical Association provided travel funds through a Beveridge Award. The University of Texas at El Paso, to which I have been affiliated since 1988, generously provided two grants that proved invaluable in coming to grips with the activity in California. I surely would not claim that I have it right, but without a travel grant in the fall of 1992, I most assuredly would have it wrong.

Historians depend on research libraries for the lion's share of their information. My gratitude goes to the staff of the University of Texas at El Paso for their courtesy and assistance. David Larkin, in particular, merits praise for his patience in leading me through a maze of government documents. Staffs at the American Heritage Center (Laramie, Wyoming), the Panhandle-Plains Historical Museum (Canyon, Texas) the Solar Energy Research Institute (now the National Renewable Energy Laboratory, Golden, Colorado) the California Institute of Technology, and the Huntington Library (San Marino, California) were generous with their time in tracking esoteric material. The archivists at Case Western Reserve Library, Cleveland, Ohio, and the Western Reserve Historical Society aided me in locating material on Charles Brush. A special thanks is due to the helpful librarians at the Engineering Library in New York City and the National Agriculture Library in Maryland. I thank them all, as well as those librarians and volunteers who took the trouble to compile files of clippings on various local topics. Their efforts with wind-energy issues—specifically at Palm Springs, Bakersfield, and Livermore, California—provided me with primary material that would otherwise have been unavailable without months of digging.

Parts of this book have appeared in a somewhat different form in *Annals of Wyoming, California History, Montana, the Magazine of Western History*, and *American Heritage of Invention and Technology*, to which acknowledgement is due.

Finally, I wish to express my appreciation to the staff of the University of Oklahoma Press. John Drayton liked the idea from the beginning and gave constant encouragement. Dennis Marshall, my copy editor, gave the manuscript a rigorous reading, improving the book not only through his many stylistic changes but also by his organizational suggestions. Sarah Iselin deserves an accolade for her patience and encouragement, and her understanding of how all these words and pictures would come together as a book.

PART ONE

EARLY HISTORY

CHAPTER 1

HUMAN USE OF WIND ENERGY

The wild free wind
They have sought to bind
And make it labor like all other things;
Nought careth he;
Joyful he works, while joyfully sings,
and wanders free.[1]

—CAROLINE TAPPAN, 1841

THE WORK OF THE WIND COMMENCED LONG BEFORE HUMAN BEINGS appeared on this earth. It began, we might surmise, when the sun began to radiate on earth, creating atmosphere, water, vegetation and other life-forms—and wind. The wind represents a primary force, created by the earth's variations in temperature and air pressure. It is a form of solar energy, generated when sun-heated air rises and cooler air rushes in to fill the vacuum: hence, wind. Scientists estimate that some 2 percent of sunlight energy received by the earth is converted to the kinetic energy of the winds.[2]

Through aeons of unrecorded time, wind had its way with the surface of the land. Erosion, which is responsible for much of the land's appearance, has been attributed more to water and ice, but wind also plays a vital role. In the American West, where natural forces often appear more evident than elsewhere in the country, weathered, mushroom-shaped granite boulders attest to the wind's never-ending work. Immense concave valleys are the work of the persistent westerlies, unremitting in their onslaught on the landscape, such as on the Laramie Plains of Wyoming. Perhaps these excavations were not so spectacular in their execution as those molded by ice, yet they are no less important. The wind, in the opinion of one geologist, has never received full credit for this "exhumation of the Rockies."[3] One can see similar wind action in numerous regions worldwide.

From the beginnings of human existence, people realized life-giving power in the wind. The thoughts of the world's early inhabitants are lost to us, but we do know that the wind figured in ritual and religion. Ethnologist Washington Matthews, who collected histories and legends among nineteenth-century American Indians, has documented some of this. He tells of an account given by a Navajo holy man in the 1890s that attributed the creation of the first human beings to the wind, Ni'lch'i. For this shaman, the wind was central. "It is the wind that comes out of our mouths now that gives us life," he explained. Continuing his thoughts in a whisper to Matthews, the old Navajo confided: when the wind "ceases to blow we die."[4] Elsdon Best related similar primal myths among the Maori of New Zealand: the "four supports that were employed by Tane [a powerful god] to support the sky are . . . the four winds" and "these winds are our salvation, were it not for them we should have no air, the breath of life would be lacking."[5]

This equation of the wind with the very breath of life was natural for humans. Wind is the motion of air, something we must constantly use. It is a life-affirming force—a continuous source of the oxygen needed for survival. For the Navajo and the Maori, the wind critically affected daily life, influencing the placement of hogans, seafaring, and hunting strategies. The wind's whimsical nature could play havoc with ceremonial dress; more significantly, it could turn a welcome rain into a crop-destroying event that spelled disaster.

Creation legends and beliefs involving the wind were widespread. The Christian Bible speaks of "an awesome wind sweeping over the world" at the time of creation.[6] The wind represented a natural power equal if not greater than other forces such as earthquakes or fires.[7] The wind could provide an analogy for a force that could not be seen, such as "the spirit of God."[8] It could also convey spiritual power, as when the people gathered for the Day of Pentecost experienced "a sound . . . from heaven like the rush of a mighty wind" that brought the Holy Ghost among them.[9]

For the Greeks, the wind occupied a central position of great potency. In their creation myth, when the goddess Eurynome, goddess of all things, danced to the south she set in motion the North Wind. Suddenly she wheeled about and grasped the North Wind. The wind grew lustful, turned into the serpent Ophion, and coupled with her. Soon Eurynome laid the Universal Egg and when it hatched "out tumbled all things that exist, her children; sun, moon, planets, stars, the earth with its mountains and rivers, its trees, herbs, and living creatures."[10] In another greek story, the god Aeolus, keeper of the wind, could use his power to assist or hamper man's progress. Odysseus, the Greek wanderer of the Ionian and Aegean seas, found this to be true when, after angering Aeolus, he and his crew were forced to row their ship on calm seas for six days. "No breeze, no help in sight, by our own folly—six indistinguishable nights and days," lamented Odysseus.[11]

The story of Aeolus and Odysseus, illustrating as it does what humans see as the capriciousness of the wind, also serves to remind us that wind power has long been used to propel ships. Just when it was that some creative individual thought to employ the wind to drive a boat is unknown—as lost in antiquity as the identity of the inventor of the wheel. However, speculation leads us to Melanesia. Some forty thousand years ago, Asians migrated to Greater Australia, following "voyaging corridors" that were relatively safe—no more than one hundred kilometers between land points. No artifactual proof exists that they used sails, but anthropologist Geoffrey Irwin believes they had substantial, though unsophisticated, sailing craft.[12] Much later, around 1500 B.C., these peoples began astounding sailing feats. Using stable, seaworthy, and fast sailing canoes, forty to fifty feet in length, they colonized the many islands of Polynesia. In their double-hulled craft, they performed remarkable navigational and sailing feats. Certainly no other contemporary civilization possessed such knowledge of winds and deepwater navigational techniques.[13] In the Mediterranean region, water transport was not nearly so sophisticated. However, it is generally conceded that the Egyptians—a river people—were propelling small craft equipped with sails of linen and/or papyrus as early a 3100 B.C. Using the current to float northward down the Nile, they took advantage of the prevailing northern winds to sail back up.[14]

In the centuries that followed, sailing vessels revolutionized travel, employing the wind to expand cultures and nations. Greek and Roman culture expanded throughout the region around the Mediterranean Sea and beyond. This expansion was based on wind power.[15] Subsequently, in modern history, there was extensive development and revision of sailing techniques, involving navigational instruments, sail making, and ships. Exploitation of the wind allowed the enlargement of the European world—a period that historian Walter Prescott Webb described as the frontier that lasted four hundred years.[16] Discovery, conquest, and trade all relied on sailing ships. From ancient times to the middle of the nineteenth century, in the bubbling wakes and billowing sails of thousands of ships we trace the paths of commerce and conquest. With the rare exception of oared boats—little used for long voyages—there was no other power source available.

The first terrestrial use of the wind again is somewhat conjectural, a matter of speculation rather than fact. When did the first windmill turn in the wind and begin the centuries-long harvest of wind power? "The quest for such hidden knowledge is beset with false trails and episodic detours," stated one historian who recently explored the question.[17] The idea may have come from Hero of Alexandria, a Greek naturalist who, in the first century of the common era, constructed a windmill model, then connected it to an air pump to play an organ.[18] However, Hero never produced a full-size machine, thus his invention can best be described as a toy. In truth, the early Greeks and Romans paid scant attention to labor-saving devices, the plausible explanation being that the abundance of slaves precluded the need for such technology.[19]

If we discount Hero's miniature machine, we must advance nine hundred years and look to the Middle East to find evidence of the first working windmills. But first we must note that legend has it that in A.D. 644, the Caliph Omar of Persia commanded his carpenter slave, Abu Lu'lu'a, to construct a windmill. The slave, however, rather than attempt construction, assassinated the caliph; thus, lacking evidence that the windmill was built, the story must be considered folklore.[20] For the tenth century, we have material proof that windmills were turning in the blustery Seistan region of Persia. These primitive, vertical carousel-type mills utilized the wind to grind corn and to raise water from streams to irrigate gardens. Although rudimentary, they were ingenious; their use soon spread to India, other parts of the Muslim world, and China, where farmers employed them to pump water, grind grain, and crush sugarcane.[21]

ENTER THE POST-MILL

The free benefit of the wind ought not be denied to any man.

—HERBERT OF BURY, SUFFOLK, CIRCA 1180

Of course, when Europeans and North Americans visualize a windmill they think of a post-mill, not the Persian carousel style.

Second only to the waterwheel in power production, the European post-mill (of which, a brief mechanical description is given later in this chapter) profoundly affected European development and society between the twelfth and the nineteenth centuries.[22] Until recently, historians of technology assumed that the European-style mill represented an East/West transfer, originating with the Christian crusaders and their observation of the Persian structures. However, the two apparatuses are very different. Aside from the fact that they both exploit the wind, the similarities are few indeed. Today the evidence suggests that European/English engineers developed the post-mill independently, hence, we have a case of multiple invention, rather than diffusion from a single source.[23]

For the European windmill, certainly the major inspiration came from the field of water power—hydraulic energy converters. The vertical waterwheel, the mainstay of what John Burke called the Medieval Industrial Revolution, provided the "parent technology" for the windmill.[24] Historian Terry Reynolds argues that "very likely the invention of the Western vertical windmill was inspired by the widely used vertical watermill, for the Western windmill adopted the vertical waterwheel's horizontal axle, gearing and machinery."[25] Surely, the waterwheel proved the inspiration to utilize another form of natural, renewable energy: the ocean tides. Saxons evidently constructed tide mills at the port of Dover, for they existed at the time of William the Conqueror. By 1600, eighty-nine tide mills had been constructed in England. On the Continent, tide mills on the Adour, a river in France, operated as early as 1125.[26]

Had the waterwheel been adaptable and versatile in all environments, there would have been no need for the less efficient, less reliable windmill. Of course, this was not the case. As Lynn White observed, the current of some rivers moved too sluggishly to turn a wheel forcefully.[27] In more arid regions, such as those bordering the Mediterranean, water might be insufficient, abundant for a few months but scant during the remainder of the year. Aridity certainly encouraged the use of windmills in the Tarragona region of Spain from the tenth century onward.[28] Furthermore, in northern countries and sites of higher elevation, winter ice could easily paralyze a waterwheel for weeks and possibly months. In addition, social and economic considerations played a part: cities and congested waterway areas welcomed windmills because hydraulic sites were overexploited. French historians have noted that in rivers where navigation was extensive "the installation of every new water-mill caused an uproar from other users of the same waterway."[29] Windmills could and often did resolve the dilemma. Although the waterwheel spearheaded the switch from human toil to mechanical labor, the Western vertical windmill found its place, too. Proof of the windmill's value may be seen in its almost explosive spread throughout Europe. Windmills attained such popularity that the pope attempted to tax the new novelty.[30]

Environmental factors and the inventive medieval mind partially explain the origins of windmill technology. However, there were also other significant influences. Historian Edward Kealey argues convincingly that the English middle classes developed and disseminated the vertical windmill in the twelfth century. Why would England, a land of rainfall and rivers, provide inspiration for wind machines? Why not simply rely on the waterwheel? The answers lie in the political circumstances of English society; more specifically, they may be found in the ownership and control of the waterways. The king of England and the Roman Catholic pope vested general riparian water rights in a few men and women of nobility and wealth and in the church. "The use of the manor's waterwheel as energy," stated historian Carolyn Merchant, "was controlled by the lord, who raised revenue from its use by the villagers and tenant farmers."[31] To protect these inherited monopolies on grinding grain, authorities rarely granted new water rights. The baronial magnates and religious institutions all jealously guarded their rights and privileges, and the thousands of water mills exacted a steady income as well as a measure of social control over the peasants.[32] It was a closed society, somewhat similar to the "hydraulic societies" of ancient China, where political power was vested in ecological control. The entrenched oligarchy partially ruled the peasants by controlling the water.[33] Twelfth-century England was, perhaps, more moderate in its repression through technology, but such control did exist.

This monopolistic water-power world was the scene for the introduction of the English post-windmill. The first one was erected in A.D. 1137 by William of Almoner, of Leicester. A novel invention, it "upset this careful balance," argues Edward Kealey in *Harvesting the Air*. New entrepreneurs realized that the wind was free and the landed aristocracy could not control this power source. Furthermore, a windmill could be erected with modest capital. Thus, "brash independents—clever peasants, knights dissatisfied with their small fiefs, university-trained intellectuals, and women anxious to support themselves—soon built their own rival mills." As can been seen the typical windmill operator belonged to the middle class.[34]

In the century that followed, these agrarian small-business people constructed numerous post-mills throughout the English countryside, causing no end of acrimonious and litigious dispute. Although it seems obvious that the air in natural motion represents a power source free to all, the nobility of that day, with the audacity reserved for the privileged, contested the fact.

One cleric proved particularly tyrannical. Abbot Samson provides a most colorful example of a religious ruler who considered himself omnipotent in the physical world: his power befit his name. Samson ruled a monastery, Bury Saint Edmunds, in Suffolk, from 1182 to 1211. On the outskirts of Bury, some three-quarters of a mile from the abbey church, a rather aged rural dean named Herbert erected a post-mill on his own land. Hearing of Herbert's competing mill, the red-bearded Samson flew into an almost crazed rage and

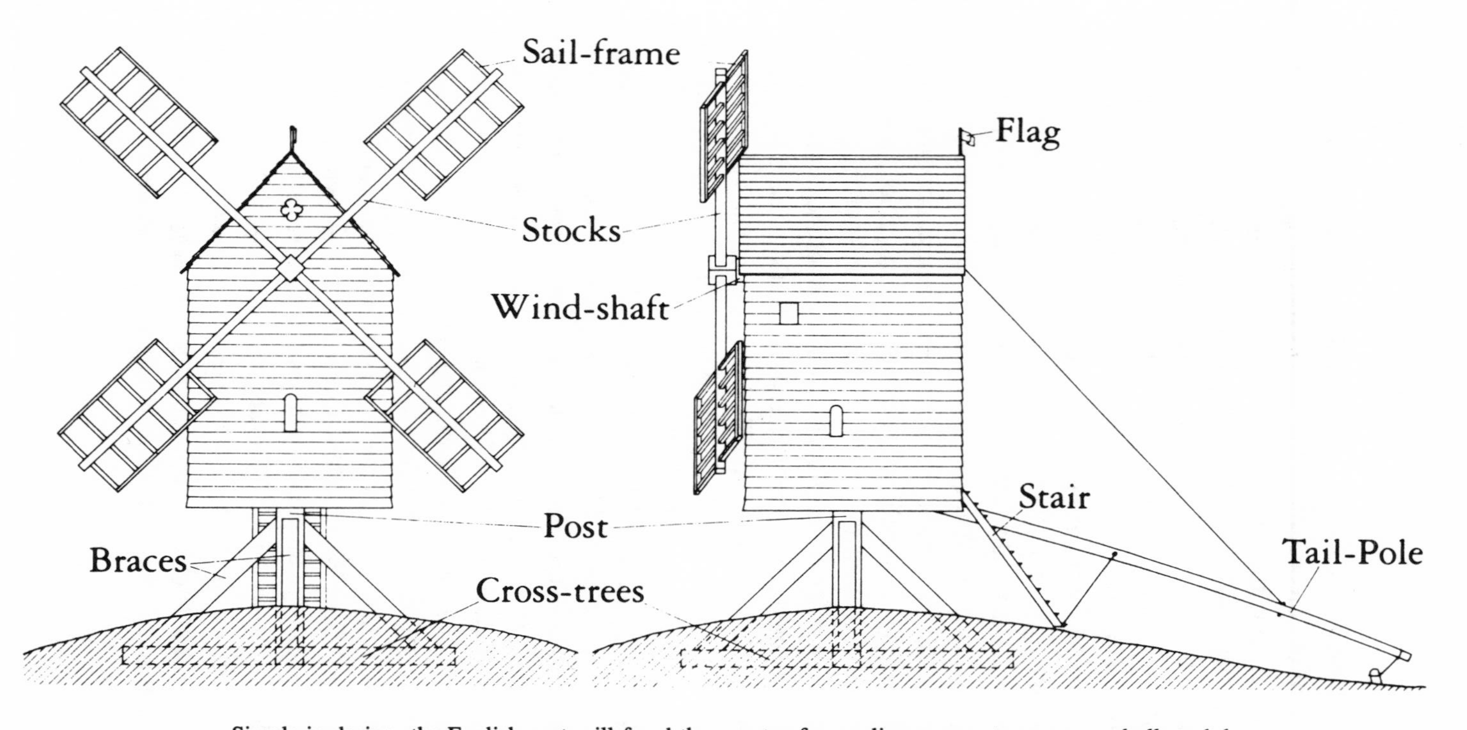

Simple in design, the English post-mill freed the country from reliance on water power and allowed the middle class to participate in energy production. (Courtesy of Terence Paul Smith, from his drawing in *History Today*)

ordered his carpenters to tear down the offending windmill. Once advised of the prospect for the mill, Herbert confronted Abbot Samson, arguing that "the free benefit of the wind ought not be denied to any man." He attempted to conciliate the dictatorial abbot, promising to grind only his own grain, thus foreswearing any intention to infringe on the abbot's monopoly. But all arguments and concessions were to no avail. The worldly religious Samson maintained that he and Bury Saint Edmunds retained the privilege of determining what could and could not be done in the environment. He did not claim control of the wind, but he did reserve the right to determine land-use zoning—that is, what would be built or not built in the region. He would not countenance Herbert's promise that the new mill would offer no competition, perhaps fearing the precedent that such a liberal concession might set. Samson again ordered that Herbert's mill be demolished, but before his workers could accomplish the deed, Herbert's sons and servants dismantled the offending windmill. In the years to follow, however, windmills multiplied, but always in areas outside of town and only under the abbot's license.[35]

Across the English Channel, in France, the first windmill had made its appearence in 1180 on the Cotentin Peninsula. A decade later, two more were constructed near the Mont-Saint-Michel and at the mouth of the Somme. These windmills not only converted energy but acted as "a breath of freedom." French historians Jean-Claude Debeir, Jean-Paul Deleage, and Daniel Hemery have noted that "it is not by chance that the period of their [windmills'] spread was precisely that when protests against lordly monopolies was [*sic*] rising." Certainly, windmills were constructed when useful hydraulic sites were used up, but they also had political and social meaning. "The windmill," maintain these historians, "was often imposed against the lord's monopolies and privileges: the new machine was generally built in opposition to the feudal lord, and its builders sometimes found allies among the royal retainers." Thus, on the Continent were windmills "established in the conditions of freedom that opened with the growth of cities, and established a further breach in the lords' energy monopolies."[36]

Generally in the late twelfth century, inherited privilege triumphed over the progressive influence of a new technology. Although some nobles did have the audacity to claim regulatory power over the wind, most suppressed competition through control of water rights and/or zoning power over land within their jurisdictional hegemony. Nevertheless, windmills proved particularly alarming, because they utilized a primary power source that the lords could not oversee. A tiny ruling class, which thrived on monopoly and privilege, began to lose out to rural entrepreneurs and a growing urban middle class.[37]

Some members of any society will always oppose technological change, and often for compelling social reasons. Others are motivated by self-interest. The English aristocracy now and then found allies at the other end of the social ladder. Occasionally, laborers joined in opposition to the spread of wind power. In 1768, for example, workers in the sawmill town of Limehouse, near London, protested the construction of a windmill. Fearing the loss of their jobs when their complaints went unanswered, they destroyed the windmill.[38] Such action might slow the spread of the windmill, but a technological advance that benefits humanity cannot be stopped. The use of the post-mill spread throughout England and Europe, generally in a west-to-east direction. By 1300, windmills could be found in Spain, France, Belgium, the Netherlands, Denmark, the German principalities, and the Italian states.

THE TOWER-MILL

As the windmill spread and multiplied, engineers created more sophisticated variations of the post-mill. For our purpose, it is enough to say that as they multiplied in number they grew in size. The economy of scale proved applicable. To accommodate larger windmills, mechanical engineers designed the tower-mill. Unlike the post-mill (in which the whole tower and mechanism turned to the wind—hence, *post*-mill), with the tower-mill only the sails, windshaft, and brake-wheel rotated to face the wind. The Dutch widely adopted this design and it was estimated that, at the height of its popularity, ten thousand tower-mills were in place and operating. In Zaandam alone, a suburb of Amsterdam, in the late seventeenth century from eight hundred to one thousand tower-mills served the town's various industrial activities. Whether by post-mill or tower-mill, utilization of the wind was a marvel, truly "a triumph of mechanical engineering and the most complex power device of mediaeval times, even up to the beginning of the Industrial Revolution."[39]

The person in charge of one of these machines bore considerable responsibility. Historian John Reynolds compared the miller to the master of a sailing ship: both depended upon the wind for their livelihood, yet each realized that its power could destroy that very livelihood. High winds often proved disastrous. Although designers had anticipated most calamities, the European-style windmill lacked an automatic reefing mechanism that would come into play when the wind quickened. To adjust the canvas sails, the miller had to bring the rotor to a standstill. As on the high seas, playing a guessing game with the god Aeolus sometimes courted disaster. "If he [the miller] hung on to his canvas too long in a rising wind," wrote Reynolds, "the mill might 'run away' beyond control of the brake, and burn to the ground before his eyes."[40] The seat of the danger was in the friction between the grindstones. If the miller had sufficient grain to buffer the stones, disaster could be averted; if not, a gale could be as destructive to a windmill as to a sailing ship.

Like ships, windmills were perceived as being of feminine gender, and many of the machines in Holland bore women's names. There is no precise explanation for the practice, but perhaps it involved their power of production or renewal, as well as the power of the miller (an occupation that was restricted to males) to control the windmill.[41] Another quality of the windmill (one not noticed by Don Quixote and the less observant) resided in its aesthetic appeal. The Dutch windmills were and are attractive. They were often objects of architectural innovation, being given wonderful wood decorations and colorful paint. The windmill was the artistic centerpiece of many a village. They have, of course, largely disappeared, being counted among the victims of steam, coal, and petroleum. Not only did the local windmill provide artistic relief from the drab, it often served as the place of public gatherings, wakes, and celebrations. "During the working day an upright cross might be a sign of mourning," explained John Reynolds, "and in some parts of [Britain] a complicated code involving the removal of shutters was employed to indicate a death in the miller's family." The stationary position of the sails might indicate mourning or a celebration, and at times of community rejoicing the windmill was dressed in "brightly coloured flags and garlands."[42]

But the purpose of these majestic windmills was, of course, work: they were machines of power and utility. They pumped water and ground corn and wheat. Entrepreneurs also employed variously designed post-mills and tower-mills for other purposes. They reduced pepper, and other spices, cocoa, dyes, chalk, and paint pigments. Lumber companies employed them as primary power for sawmills. Paper companies used windmills to reduce wood pulp to paper. Although not as versatile as today's machines, in many ways these striking windmills provided rotary power similar to twentieth-century electric motors.[43] How much energy did the windmills actually provide? It is impossible to estimate with precision, but one authority has calculated that wind machines supplied as much as 25 percent of Europe's industrial energy from 1300 to the coming of the steam engine and cheap coal in the 1800s.[44] The remainder came from hydropower, assisted by human and animal labor.

Dutch windmills combined utility with beauty of design. Often they incorporated architectural innovation. (From Laeendert van Natrus, et al., *Groot Volkomen Moolenboek*, vol. 1, plate 8. Courtesty the Huntington Library, San Marino, Calif.)

TUMULT IN THE AIR—AND IN SOCIETY

As well as having social consequences, some of which have already been mentioned, the wind-energy advances of medieval and Renaissance Europe produced major technological progress. To a degree, the early carousel windmills of Persia and the Middle East had encouraged technological advance, but technology today is more beholden to the English post-mill and the Dutch tower-mill, with their horizontal axes. This period produced many seminal ideas: one need only note the dominance of the horizontal wind turbine, relegating the vertical style to a secondary position. Invention, like discovery, magnifies its value through replication and repetition. Today, Americans still rely on basic engineering and design principles developed by the technological wizardry of creative medieval minds.

Furthermore, the Dutch first gave notice of what a number of windmills working in concert could accomplish. Perhaps the first of what have come to be known as wind farms were the horizontal-axis windmills working together to drain the polders and create Holland's rich agricultural land. In reclaiming the land from the sea, hundreds of Dutch windmills worked collectively to alter nature. Lewis Mumford wrote of this effort that

> the gain in energy through using wind and water power was not merely direct. The mammoth machines made possible the restoration and cultivation of rich soil, reversing a historic degradation. . . . Land building and irrigation, are signs of a planned, regenerative agriculture: the windmill added absolutely to the amount of energy available by helping to throw open these rich lands, as well as protecting them and helping to work up their ultimate produces.[45]

Wind energy accomplished an environmental task that neither human or animal effort could have achieved.

But we must return to how the development of windmills prompted social change. In the monarchial and ecclesiastical world of the Middle Ages, only the rich and the well-born possessed individual rights. Such rights and powers were a matter of inheritance rather than initiative or accomplishment. Classes were static; social mobility was an untried concept. In a sense, the windmill proved a stimulus, perhaps a catalyst, to budding ideas of middle-class enterprise and democracy. In the same era that King John capitulated, conceding the political rights expressed in the Magna Carta, wind machines illuminated new entrepreneurial possibilities. The wind and the sun are, perhaps, the only primary power sources that cannot be controlled. Portions might be captured, but neither energy source can be monopolized. Therefore, the privileged class, represented by the likes of the aforementioned Abbot Samson, found the wind as socially destructive as it could be physically tumultuous. Enterprising men and women, seeking economic independence and a degree of political rights, built windmills in defiance of established energy monopolies and feudalistic practices. They were not altogether successful, but this mechanical marvel introduced one avenue to begin questioning the long-established medieval order. As is often the case, technological advance proved to be the harbinger of social change.

Perhaps the most significant role that the windmill played was in advancing the industrial revolution. Although historians ascribe this revolution to the late eighteenth century, Lewis Mumford argued that "if power machinery be a criterion, the modern industrial revolution began in the twelfth century and was in full swing by the fifteenth." "The greatest technical progress," he continued, "came about in regions that had abundant supplies of wind and water."[46] These "nonautonomous machines multiplied power" through use of the natural forces of wind and water.[47]

This ability to create energy—to multiply power—from natural sources changed social and economic relationships. The wealthy gained leisure time; the poor were freed from servitude. "Thanks to the menial services of wind and water," asserted Mumford, "a large intelligentsia could come into existence, and great works of art and scholarship and science and engineering could be created without recourse to slavery: a release of energy, a victory for the human spirit."[48] Historian Lynn White seconded this appraisal of the social importance of technology: "The chief glory of the later Middle Ages was not its cathedrals or its epics or its scholasticism: it was the building for the first time in history of a complex civilization which rested not on the backs of sweating slaves or coolies but primarily on non-human power."[49] Spinning windmills harnessed power, and in the process they freed human beings.

WIND IN THE COLONIES

As European countries explored and expanded to the New World, we might ask, did the windmill accompany them? The answer is yes; but it is a very limited affirmative. The windmill cannot be counted in that bushel basket of cultural traits and technological devices that successfully crossed the Atlantic Ocean with the invading Europeans. It did make an appearance, but for a number of reasons the windmill did not fare well. Attempts to transfer feudalism to the American colonies failed, thus the social and political stimuli operating to the benefit of wind technology in England and France were not present in America. As a technology, the European-style windmill (particularly as developed in Holland) was large, cumbersome, and relatively expensive. Once constructed, it could not be moved. Furthermore, the European windmill required constant human attention. In a colonial world of abundant land but few settlers, labor was scarce and therefore valuable. Any apparatus that required continual attention would be at a disadvantage in the panoply of American technology.

Environmental conditions proved equally important. Colonial America boasted fertile soil, plenty of rainfall, and free-flowing rivers, but, except for the coastal areas and mountaintops, not much wind. Quite naturally, colonists concluded that the watermill should be the power source of choice. One historian has noted that "throughout rural America . . . simple but effective country mills, driven in most instances by water, nearly everywhere followed closely on the heels of settlement and persisted long after the days of pioneering had passed."[50] Few communities were without water-powered mills for long. Food and shelter were the first orders of business in building a community. To provide such necessities, corn must be ground and logs must be sawed. Colonists could best accomplish these tasks through the construction and use of a waterwheel.[51]

In the first half of the nineteenth century, waterwheels multiplied. Although opinions varied, a figure of 7,500 is reasonable for the first census year of 1790. The tally increased to some 55,000 by 1840. Generally, the horsepower of most watermills was not great, but as an inexpensive, renewable form of energy, hydropower merited the attention of industrialists and pioneer farmers alike.[52]

Although the waterwheel was well suited to the American environment, the social consequences of its introduction were diverse. In the northern colonies, the waterwheel assisted in the development of small businesses and the capitalist system, decreasing the need for backbreaking human labor or animal toil. In the southern colonies, however, waterwheels contributed to a failed feudalistic order that was being supplanted by a slave-holding society. As in England, the capturing of energy from water lent itself to monopoly and antiegalitarian tendencies, simply because it was concentrated (and therefore manageable), rather than diffuse. Historian Larry Hasse maintains that the waterwheels of the South served well the planter aristocracy intent on maintaining and nurturing a slave-labor system.[53]

Although the European windmill floundered in the colonies, a victim of New World environmental realities, it did find limited use. Soon after the 1607 establishment of Jamestown, the Virginia Company directors instructed Governor George Yeardley to construct water mills, horse mills, and windmills. Yeardley, following orders, in 1621 erected the first windmill in the colonies on his Flowerdew Hundred plantation.[54] Just how many windmills resulted from the Virginia Company edict is unknown, but we do know that William Robertson constructed an English post-mill near Williamsburg about 1720.[55]

Windmills of both the post and tower designs found considerable use on islands and coastal spits. Not surprisingly, in New York City (New Amsterdam) a number of tower windmills were built by the Dutch founders. In nearby Rhode Island, English settlers constructed a windmill in 1639, shortly after their arrival. Long Island, particularly Cape Cod, afforded the windy conditions necessary for successful windmill use, and in the seventeenth and eighteenth centuries settlers constructed grain-grinding machines. In hilly, wooded regions, loggers erected windmills when water supplies were inconvenient. One historian noted that "in many instances, windmills were used to drive the sawmill" in New England.[56] Smock windmills—so named because when shingled and whitewashed they resembled a rural women's frock or smock—dominated. Most residents found them attractive and colorful, although the irascible Henry David Thoreau thought they resembled huge turtles.

A few early Americans predicted a role for windmills in the new nation's industrial future. Trench Coxe, an ambitious Philadelphia merchant and perceptive economic theorist, implored his friend Alexander Hamilton and other Founding Fathers to write a strong federal constitution to encourage industrial self-sufficiency, noting that the new nation "by wind and water machines . . . can make pig and bar iron, nail rods, tire, sheet-iron, sheet-copper, sheet-brass, anchors, meal of all kinds, gunpowder."[57] Although this projection for wind energy went unheeded, most Eastern cities did come to boast at least one European-style windmill, and in the West—as will be noted below—San Francisco built two.

However, windmills of the European variety were the exception rather than the rule—perhaps cultural reminders of the Old World. They were not essential to the New. In the colonial period and the decades leading up to the Civil War, Americans created energy through human and animal labor, wood, and water. People employed wind, but primarily at sea. The terrestial importance of wind energy changed rather dramatically, however, in the later half of the nineteenth century. Ironically, at the very time Americans replaced wood and water with steam and coal, energy plucked from the restless air found a nitch in the energy mix. The place was the American West and the wind-harvesting invention became known as the American windmill.

THE AMERICAN WINDMILL

Much of the West is arid or semiarid. It is a land of little rain but frequent wind. The god Aeolus is ever present, cavorting continually on the plains, the valleys and the mountaintops of this vast region. The wind is blamed for erosion and destruction of property—and for human suffering in the form of injury or death from powerful storms, psychological depression, and even insanity.[58] A summer rain can turn into a tornado. Westerners see and hear winds knock down trees, flatten gardens, blow off roofs, and carry off most everything that is not battened down. Wind in the West creates sandstorms of destructive capabilities; on Western lakes it raises waves of fearful proportions. It is continually a topic of conversation, but rarely is a good word spoken for wind, save its ability to clear the air during the mosquito season. In the 1930s, the western wind was known as a malicious force that contributed to the Dust Bowl as it sifted the topsoil from the plowed plains, swirled it skyward, carried the soil across the continent in a great dust cloud, to deposit it where it wasn't needed. Many Westerners would agree with novelist Owen Wister's opinion, stated in a letter to his mother in 1885: "One thing about this country I don't like—and only one. The wind."[59]

Yet this vexatious agent performs valuable services. It may remove soil, but it also transports moisture. The wind assists plants with the task of procreation through pollination and the spread of seeds. In the nineteenth century, winds occasionally assisted westering Americans in an unusual fashion. Although "wind wagons" fall somewhere between myth and reality in Western history, the harnessing of wind to roll westward did enjoy a brief existence. Early travelers on the Great Plains often found the ocean a convenient analogy to describe the treeless, expansive, and often level land, and since pioneers had sailed across the oceans of the world, why not sail on the prairie?[60]

Samuel Peppard constructed the best-known wind wagon in 1860. Peppard was an inventor with a yen to visit the Colorado goldfields. To get there he built a 350-pound contraption shaped like a boat, with wheels, and rigged with two sails. Amazingly, Peppard and two companions sailed across Kansas to a point near present-day Fort Morgan, Colorado, a distance of some five hundred miles. At that point, a whirlwind caught the wagon, lifting it twenty feet into the air before unceremoniously dumping it on the baked prairie. The rear axle being fractured, Peppard and his mates hitched a ride to Denver, leaving his wagon to its desert fate.[61]

Although the wind wagon provides a comical footnote in Western history, wind as the force behind water-pumping windmills proved quite pivotal in the settling of the West. Such water pumps assisted the early pioneers and they are still in use today. According to historian Walter Prescott Webb, it was the American windmill, in concert with barbed wire and the six-shooter, that allowed for successful settlement of many of the more parched places in the West.[62] This windmill represented intermediate technology at its best. It transformed the abundant wind into an agent to alleviate the shortage of water. It liberated groundwater for a moisture-poor region, providing the technology necessary to settle vast tracts of the rangeland, creating stock wells and small, vernal oases.

The American windmill design bore little relationship to its European predecessor: it was small, multibladed, light, movable, self-regulating, and easy to maintain. It was also inexpensive. Manufacturers mass-produced various models and the parts were interchangeable. Ranchers and farmers used these windmills mostly to pump water, although a few of them expanded into grinding grain and doing other rotary-type tasks. In the broad sense, the new windmill significantly contributed to farm mechanization—the replacement of human energy by machine power.

Although it was the Western environment that created the demand for a new windmill, a Connecticut mechanic, by the name of Daniel Halladay, provided the inventive genius. In 1857, Halladay, having perfected his windmill, formed the Halladay Wind Mill Company.[63] Moving west, Halladay and his partners later incorporated in Batavia, Illinois, as the U.S. Engine and Power Company. The company struggled, but by the mid-1870s it was financially successful.

It was Western railroad builders who first used the Halladay windmill. Hand in hand with the first transcontinental railroad came the windmill, providing water to the thirsty Union Pacific steam locomotives. The company installed windmills of various sizes at watering places along the line. The largest boasted a wheel with a diameter of thirty-nine feet. Halladay built these machines to withstand the gale-force winds that swept across the southern Wyoming and Nebraska plains, protecting the structure by having the blades in sections (eight to a section), attached to the large wheel. When the wind grew in force, each section furled inward,

The Halladay was the first commercially produced American windmill. This model, at thirty-nine feet, was purchased by the Union Pacific Railroad and installed at Laramie, Wyoming Territory. (Courtesy American Heritage Center, University of Wyoming; and Union Pacific Railroad Archives)

giving the appearance of an open basket, affording little resistance to air, which could pass through without causing damage.

In the years to follow, a bewildering number of styles and brands of self-governing windmills appeared. Water-pumping windmills dotted the U.S. landscape. There is no way accurately to estimate their numbers, but some authorities have offered a figure of more than six million.[65] A traveler on the Great Plains, or in the Rocky

Mountain region, the Great Basin, or the Pacific Slope, could not venture far without encountering the American windmill. Halladay, Elipse and Aermotor models and a whole variety of other brands added a vertical dimension to an otherwise horizontal landscape.[65]

These windmills provided the technology necessary for the expansion of the cattle industry. Until their introduction, a rancher's grazing was limited by surface water: a cow will not amble more than fifteen miles a day for water.[66] Thousands of square miles were unusable for livestock-industry purposes. However, by the early twentieth century, the windmill, with its adjacent stock pond, had become a familiar part of the ranching West. For many Westerners, it provided a colorful landmark on an otherwise monotonous terrain. The windmill offered another symbol of man's technological conquest of an exacting and unyielding expanse of land—although for cowboys, who were the ones who had to climb the tower to oil the machine and make repairs, it was also the source of an added chore.

Often the windmill offered not only water—for personal use, as well as for cattle—but also it was often the only landmark for many miles; therefore, both cows and cowboys congregated nearby. During harsh weather, it could be a point of reference on an otherwise undifferentiated landscape. The windmill's presence has also endured: even in today's ranching operations, which often feature such myth-destroying technology as computers and helicopters, in isolated, arid regions, the reliable windmill is still a common sight.

Although most manufacturers designed American windmills to pump water, power windmills proved a precursor to electricity-generating wind machines. They were so-named because they did more than pump water. Geared and fitted to execute stationary rotary tasks, they rendered direct power, used to grind grain, saw wood, churn butter, and perform other farm tasks. Occasionally, workers employed them creatively. A letter to the *Scientific American* in 1883 described a unique wind-powered system used by miners. A windmill raised sand by means of a conveyer belt to a large storage container; the operator then released the sand to provide power to run six *arrastres*, a spinning apparatus used to break down silver-bearing ore. The owner of the mine described the process:

A stream of sand [is] let out upon the overshot wheel. It revolves just as it would under the weight of a stream of water, and the arastras [sic] move steadily on at their work. When there is much wind, sand is stored up for use when calm prevails, so the arastras are never idle.[67]

Such a system, utilizing both wind and sand, seems remarkably well-suited to the Southwest. Just how extensively early miners took advantage of the system is not documented.

The windmill's importance, however, has not been restricted to its labor-saving qualities. It was also a cultural icon. Recently a writer summarized this significance: "The American windmill is more than a machine—it is a symbol, evoking different memories in each of us. What the buffalo meant to the Indian, what the horse meant to the Spaniard—that is what the windmill meant to the American settler. Survival—development—staying power."[68] One rancher, writing in an agricultural history journal in the 1960s, affirmed that the windmill is "a comfortable sight, assuring that water is available particularly during times of drought."[69] For others, the windmill represents continuity and a nostalgic but important connection with the past. Texas freelance writer Larry Hodge believes that "the windmill reminds us that the simpler way of life is not completely gone. As long as windmills exist, Texas will be a refuge where a handshake is as good as a contract, where old ways are kept alive not because they are old but because they work, and where slow but steady can still win over fast and glitzy."[70]

The windmill is not, after all, a technology that violates the sentimental rural, pastoral life that so many Americans praise but so few practice. As the nation has fallen increasingly under the

Estimates vary, but some say that approximately six million American windmills were operating in the Great Plains and the American West between 1880 and 1930. This "Currie" was typical of the windmill's functional value in cattle country. (Courtesy T. Lindsay Baker, Rio Vista, Texas)

influence of urban culture, the windmill has become a symbolic bulwark against modernism and rapid change. It represents technology, but a technology that is compatible with nature and the American sense of the bucolic. As the above-quoted historian-rancher wrote: "Never has man invented a machine that is in more beautiful harmony with the physical environment than the windmill in western Texas."[71] Certainly, it squeaked and clicked unmercifully, but the sound did not bring discord to the symmetry of natural sounds or the sense of tranquility known in an earlier age.

Although few may have listened to windmills, many looked at them. Observers of old windmills, whether European or American, found them visually satisfying. The nineteenth-century English writer William Cobbett caught that spirit when he wrote that he was delighted to count seventeen windmills on surrounding English hillsides: "They are all painted or washed white; the sails are black; it was a fine morning, the wind was brisk, and their twirling altogether added to the beauty of the scene, which . . . appeared to me the most beautiful sight of the kind that I had ever beheld."[72]

Throughout its history, the windmill has stimulated and inspired human kind's artistic nature. Many painters from Michelangelo on have found windmills an inviting subject. Writers and poets have sprinkled their works with allusions to the wind and to windmills. In reference to American windmills, one writer speculated that "there is something about the windmill that attracts the eye of photographers and artists. Perhaps this fascination can be partly explained by the fact that the vertical shape of the windmill tower often provides relief from the monotony of the almost unbroken horizon." It is more impressive and comforting than "fence posts, sagebrush, and cattle scratchers."[73]

American windmills in all their fascinating varieties are appreciated today by the public. They are historic, romantic, and artistic—"cultural symbols," similar to those that Leo Marx believes convey "a special meaning (thought and feeling) to a large number of those who share the culture."[74] They evoke an earlier bucolic time, simply because they employ technology at what we perceive to be an acceptable level of disruption to landscape and nature. Today, according to Dean Kilgore, of Dempster Industries, Beatrice, Nebraska, there are probably about one hundred thousand windmills still turning in the Midwest. Dempster and Aermotor still manufacture windmills for a limited ranching clientele, and it is an arresting fact that approximately one-half of all windmills sold today are "strictly for decoration."[75] What to ranchers in the nineteenth century represented a practical necessity is to modern Americans often a nostalgic reminder.

The tale of two Dutch-style windmills in San Francisco encapsulates this observation. Built in the 1870s to facilitate a city beautification project, in time they became tourist attractions in their own right. Civic leaders of the city by the Bay had set aside a parcel of oceanside land, to be known as Golden Gate Park. Eventually, this land would rival New York's Central park as a model urban park, but at that time it consisted of shifting sand dunes beset by constant winds. If the park was ever to be a civic attraction, it would have to be well planned; it also needed water. A famed landscape architect, Frederick Law Olmsted, provided the planning; windmills provided the water. By the turn of the century, park commissioner John McLaren had convinced the city fathers to build a Dutch mill, seventy-five feet tall, to pump water to a holding lake. The project proved so successful that a wealthy park proponent contributed funds to construct another windmill, five feet higher than the first. The two windmills were instrumental in transforming Golden Gate Park to a vernal oasis. Electric pumps soon replaced the windmills and they fell into disrepair, but in the 1970s concerned citizens, led by Eleanor Crabtree, raised $140,000 for their restoration. Today, the two monarchs can be seen, spinning to the delight of resident and tourist alike, reminders of the intermediate technology of an earlier era.[76]

Cultural geographers often use the terms *topophilia* and *technophobia* to deliniate the American public's perception of landscape.[77] The old water-pumpers represent a continuation of topophilia, or love of the land. Because they have a human constituency, their erection and continued use raises little public concern. On the other hand, the larger wind generators of today—the subject of this book—have a negative image in the eyes of many Americans. There is nothing nostalgic or romantic about them. They represent not the pioneer past but an uncertain future. The turbines evoke feelings of technophobia. They are steel and they are massive. The technology does not seem to fit in either the desert or the greenbelt of northern California's Altamont hills. Visually, they seem to rival nature rather than cooperate with it. Grouped together, they might appear to be a stationary, yet moving, army. Had Don Quixote been battling lines of turbines, history might have judged him quite sensible. Water-pumping windmills and wind machines that generate electricity do essentially the same task: that is, they both capture kinetic energy for human use. However, in the minds of many people they are different. The windmill, like the soaring birds, seems to be of nature, transforming the turbulent wind into tranquil energy. The metallic, robot-like, modern wind turbine seems—to such critics—to be more exploitive and indeed foreign to the landscape. Such descriptions may bear little resemblance to reality,[78] but symbols are significant and often rule human emotion.

Perhaps our perceptions are partially determined by the product. Windmills liberate water; the turbines generate electricity. Although there is no more natural resource than water, the harnessing of electricity represented a new phenomenon, one of the great leaps forward of science.

CHAPTER 2

A CONNECTION IS MADE: WIND INTO ELECTRICITY

[It is not] utterly chimerical to think of wind superseding coal in some places for a very important part of its present duty—that of giving light.

—WILLIAM THOMPSON (LORD KELVIN), 1881

THE WINDMILL WOULD BE MEANINGFUL IN U.S. HISTORY EVEN IF ELECTRICITY had not transformed the way in which Americans live. However, it is the marriage of the wind machine with electrical energy that holds particular promise for the future. It is this union—this nucleus—that is the focus of the story told in this book.

No one knows at what point in antiquity a human being first contemplated the mystery of lightning or the attraction between a lodestone and iron. Certainly, throughout recorded history the enigmatic force of what is now known as electricity has been recognized, but scientists did not systematically experiment with "magnetics" and "electrics" until early in the seventeenth century. Thus, the applied or practical use of electricity belongs to the modern era. Even as recently as the eighteenth century, scientists, intrigued with the newly identified power, were frustrated by their inability to bridge the theoretical and the practical. How could electricity be used? Writing to a fellow scientist in 1748, Benjamin Franklin described how he and his colonial colleagues planned an outdoor banquet in which the pièce de résistance, a turkey, would be electrocuted by means of a shock sent from one side of the Schuylkill River to the other. Franklin was encouraged to perform this bizarre demonstration in part because he was "chagrined a little that we have hitherto been able to produce nothing in the way of use to mankind."[1]

Franklin, had he lived, would have had to wait another one hundred years for electricity to have a significant impact upon society. In May 1844, Samuel Finley Breese Morse successfully transmitted a telegraph message across a distance of thirty miles on a line strung between Washington, D.C. and Baltimore.[2] The event marked the birth of instant long-distance communication, changing the way in which the Western world acted and communicated. When the transcontinental U.S. telegraph line linked East with West (1861), it signalled a triumph for technology and the death knell for that slow, romantic way of communication, the Pony Express. It also proved to be the harbinger for dramatic changes that electricity would introduce—an initial thrust in what historian Paul Valéry called "the conquest of the earth" by electricity.[3] It was during the nineteenth century that "practically all of the fundamental discoveries which form the main branches of the electrical industry were made."[4] By the 1880s, the development of artificial lighting, the telephone, and electric street railways all had a dramatic effect on practically every urban American. Each year, it seemed, creative Americans invented new and practical uses for electricity.[5] Electric energy, after all, was highly convenient and *mobile*. Two wires might take it to any point, where it could be transformed to suit the consumer's purposes whether that be electrical apparatus, motor, or light.[6]

Increased demand sparked a debate among engineers about how electricity should be produced. Electricity could be furnished by the heat energy of a boiler-furnace (fueled by wood, coal, or petroleum) or by the kinetic energy of movement, either of water or air. How should dynamos be activated—by energy from wood, coal, petroleum, the flow of water, or wind? Because the United States possessed abundant energy sources and sufficient wealth it had the luxury of choice. The nation's engineers initially cast their lot primarily with coal, later with petroleum. Both coal and petroleum were abundant and therefore inexpensive, and could readily be converted to heat and energy. Engineers also embraced hydroelectricity, although only to a limited extent.[7] Wind—a diffuse source—was not a serious contender.

Although, in the search for electricity providers, wind energy was dismissed, it is remarkable how widely midnineteenth-century America employed windpower. Sam Schurr and Bruce Netschert, two economic historians, have estimated that in 1850 windpower accounted for 1.4 billion horsepower-hours of work; waterwheels provided 0.9 billion horsepower-hours, and steam power (from bituminous coal) came in a distant third with 0.4 billion.[8] These statistics, of course, are estimates rather than hard data. It is also important to note that the predominance of wind power is based on its widespread utilization by sailing ships, more than its terrestrial use. Nevertheless, such general findings give notice that Americans recognized and utilized windpower in staggering amounts. This use of the wind continued for many decades. At the close of the nineteenth century, "harbors still contained forests of sailing vessel masts and the dominant feature of many rural landscapes was the windmill."[9]

But industrial America became fascinated by the economies of scale. Historian Howard Mumford Jones described the late nineteenth century as "the age of energy," a period when Americans became obsessed with the ability of new energy sources to transform the nation's towns, cities, and landscapes. In that age of energy boom, scientists and manufacturers willingly cast aside the more gentle, diffuse renewable energy sources for fossil fuels—the beginning of what Amory Lovins has since labeled "the hard energy path."[10]

By 1890, the share of wood, water and wind in the provision of power diminished to approximately 10 percent. The "soft" fuels had been almost totally supplanted by steam, coal, and petroleum. In Schurr and Netschert's assessment, "windpower and direct waterpower were relegated to minor positions as sources of mechanical energy."[11]

The questionable consequences of coal and petroleum use for the environment seemed scarcely to have been raised. In an era when a belching smokestack symbolized progress, few Americans were concerned that both coal and petroleum were nonrenewable resources and that both would have harmful effects on the physical environment—especially on air quality. It was an ebullient era. Few Americans fretted about modifications of the environment, and phenomenon such as "landscape guilt" and technophobia were many decades in the future.[12]

In the midst of this electrical revolution—this second industrial revolution—a few inventors and scientists questioned the direction of energy production, particularly the reliance on coal. They urged colleagues not to abandon renewable sources of energy such as solar, tidal, and wind, but rather to modify and to improve their use. Such urgings had little to do with concern about a clean environment; they were based more on ideals of frugality and common sense. Why not employ a free energy source if it was available?

An Englishman, William Thompson (Lord Kelvin), first urged such a course. In 1881, this renowned physicist stated in a lecture that in the headlong dash to industrial power, nations ought not to forget the value of natural, renewable, sun-created sources of energy. The sun, Thompson reminded fellow scientists, provided the source of all energy, save a little from the motions and attractions of the moon and planet Earth. Thompson found wind power particularly fascinating. He reminded his audience that of Britain's forty thousand ships, ten thousand were steamers and twenty thousand were sailing vessels, creating "vast amounts of horse power." When combined with the sailing ships of the world and when thrown in with "the little item of windmills," old fashioned wind still supplied a large part of the energy used by man.[13]

Thompson's purpose was not to praise the past but to warn of future dependence on coal. Already England had decimated its forests and was busy mining its coal, a finite resource. "The subterranean coalstores of the world are becoming exhausted surely, and not slowly, and the price of coal is upward bound," he warned his audience. The "lamentable decadence of wind-power" should be reversed, and therefore "it is most probable that windmills or windmotors in some form will again be in the ascendant." He did not believe it "utterly chimerical to think of wind superseding coal in some places for a very important part of its present duty—that of giving light."[14]

Thompson's enthusiasm was not due to any break through in windmill technology, but more to efficient dynamos and advances in electrical storage. This eminent scientist was particularly enamored with the new storage battery ("accumulator") developed by the Frenchman Camille Faure. Faure's accumulator, which Thompson referred to as a "box of electricity," solved the storage problem. As Thompson's biographers put it: "Sir William saw in the Faure cell a method of preventing failure of light or power should the generating machinery break down."[15]

Of course, windmills, too, might or might not break down, and their energy source often did. But doldrums of wind could now be compensated for by a store of energy. Thompson suggested that "with new dynamos and Faure's accumulator, the little want to let the thing be done is cheap windmills." With the system he proposed, the windmill would charge the battery over a six- to twelve-hour period. The drawing off of the charge (illuminating a sixty-candle bulb) would be done over two hours. If an unusual calm prevailed, "a little steam engine" could be ready, or "the user of the light may have a few candles or oil lamps in reserve."[16]

When Thompson mentioned a "little steam engine" or candles (for the less affluent), he identified a major problem with wind energy: its inconstant nature. Without an alternative primary source, wind energy could not guarantee power to a consumer when it was required. No factory, could rely on wind power without having either a prohibitively expensive storage system or a labor force on call twenty-four hours a day. Furthermore, the wind, as diffuse, kinetic energy, could not generate power comparable to the steam engine or the water turbine.

Because of the wind's capricious nature, Thompson and other scientists realized that, without an economic and effective method of storage, this form of energy was destined to the scrap heap of brilliant but impractical ideas. The storage problem would be the Achilles' heel of wind energy, but because inventors were intrigued with the obvious benefits of a free source of energy, they wrestled with the problem.

LOOKING FOR A BREEZE

It seems incomprehensible that such a ready and potent agent [the wind] should escape practical use so completely that it does.

—Scientific American, 1883

The *Scientific American*, a weekly journal committed to science, invention, and American technology, offered one clearing house for ideas. Within its pages appeared articles, anecdotal pieces, editorial comments, letters to the editors, and articles sometimes rather questionably lifted from similar journals, both domestic and foreign. Occasionally a lofty, theoretical article might appear, but the editors preferred the applied sciences. The publication acted as a sounding board not only for practicing scientists and engineers but also for designers, managers, entrepreneurs, and other Americans enamored with the possibilities and the progress of technology.[17]

Use of the wind was a frequent subject, sometimes in serious debate, sometimes in a jocular vein. A letter from Rufus Porter, age ninty-two, was in the latter category. Known by the editors for his "quaint writings and the extraordinary results he always expects from his wonderful inventions," the venerable inventor proposed a wind plow of considerable proportions. It would till four acres per hour. It would, predicted the ebullient, eternally-optimistic Porter, "harrow, sow, reap and mow, thrash grain, shell and grind corn, carry loads, irrigate land at the rate of 100 acres a day, or will travel 10 miles an hour in any direction, with 20 passengers." In a memorable understatement, he added: "But all these things require a good breeze."[18]

Most inventors envisioned a less ambitious role for wind energy. Pumping water was the function most favored, and earnest innovators were busy redesigning and perfecting the American windmill.[19] A few, however, were thinking about combining the ancient power of wind with modern electricity. The first to put this thinking into practice was Moses Farmer, who in 1860 patented a device to convert wind power into electricity.[20] An electrical pioneer, Farmer also perfected a striking device that would be useful as a fire alarm, invented an incandescent light, and came up with a self-starting dynamo. His incandescent light was too expensive to be practical. In truth, Farmer was ahead of his time and profited little from his genius.[21] In 1860, after all, what was a person to do with electricity?

With the close of the Civil War, interest in electricity quickened. One inventor, anticipating a diminished supply of the nation's coal, suggested that wind power be employed to create heat, primarily through a "magneto-electric" machine.[22] Another inventor advanced the idea that on farms wind machines could "be employed to produce a reservoir of that fluid [water] to be used on a small turbine or water motor, to give motion to a sewing machine or knitting machine, a loom, a churn, and for various other purposes."[23] In the absence of suggestions about a dynamo or a storage system, this inventor presumably had in mind the direct energy of a power windmill. In the early 1880s, interest in the potential of wind energy further increased. For six months in late 1883, the editors of the *Scientific American* published an exchange of information and viewpoints regarding wind energy. In July 1883, the journal's editors posed a central question, one of importance for the future of the nation: "How can we best turn to account the natural forces which are in play about us?" These forces, according to the journal, consisted primarily of "two fluids"—water and air. Although Americans had utilized water power, the percentage of wind power in use was "practically inappreciable" and it was time for a change. "It seems incomprehensible," the editors declared, "that such a ready and potent agent should escape practical use so completely." The journal acknowledged that wind power was rather unmanageable—"destitute of all uniformity" and "sometimes furious, sometimes absolutely nothing, and at all times unsteady and capricious." But the editors anticipated that American technology was up for the challenge.[24]

If the creation of electricity from the wind posed plenty of puzzles, an even greater challenge was how to utilize it on a large scale. How could a business or factory successfully use wind energy? "How," asked the *Scientific American* editors, "shall we *store* the power that may come to us by day or by night, Sundays and week days, gathering it at the time we do not need it and preserving it till we do? This is the problem. Who is the man to solve it."[25] Readers quickly responded. Although the editors placed some faith in accumulators or batteries, such devices were questioned on account of expense, safety, and reliability. One knowledgeable reader, who chose not to identify himself, wrote that since the life of a battery was short, "batteries [were] . . . in truth of small value." This writer favored energy storage through air pressure, but admitted that a suitable container would be beyond the means of most people. In spite of such costs (estimated at $900) the possibilities of creating air pressure with windmills interested other writers. One creative mind suggested a rowboat incorporating a small wind wheel. A two-inch pipe—installed on the port and starboard sides of the craft—once charged "will contain sufficient air to give the power of one man for five consecutive hours."[26] Another suggested use for such a wind wheel was on a "light vehicle, or buggy," the "air chest" being attached to the axle of the driving wheels.

Ground transportation was, for most Americans, more important than water travel, thus the idea of supplanting animal power with wind power and compressed air had some appeal. One reader of the *Scientific American* suggested that a light vehicle propelled by an air-cylinder "ought not to exceed the cost of a team [of horses]." He had in mind using interchangeable cylinders: while one was in use, the other would be pressurized by the wind generator. This would provide "a first class means of locomotion, whereas now comparatively few *good* horses can be found."[27] The idea may have violated the Second Law of Thermodynamics and it does not appear that a wind vehicle of this type was constructed.

In stationary applications, one of the more bizarre ideas for storage of power involved weights and springs. The windmill would be installed above a factory, with gears designed to lift a large weight, gradually, from the floor to the roof of the building. When the factory called for power, an engineer would allow the weight to settle gently to the floor by what the inventor termed "the clockwork principle," allowing for two to three days of power.[28] The editors of the *Scientific American* were skeptical. They estimated that a three-day supply of power for a small factory would require a weight of 24,000,000 pounds suspended to a height of fifty feet. Since one windmill could lift a maximum of seven tons,

1,600 weights (and 1,600 windmills) would be required. When the editors tried to estimate the strength and cost of the building to hold such a weight, they admitted: "We do not feel competent to the task; our arithmetic has given out."[29]

A storage plan involving a coiled spring was equally fanciful. A correspondent suggested that large windmills would slowly force down a huge spring "and then drive the machinery by [its] recoil." The editors acknowledged that if this storage system could be accomplished "it will make a wonderful advance." However, they warned—perhaps tongue-in-cheek—that once the spring was fully coiled with energy it will be "a fearfully dangerous instrument to manage." Both the weight and the spring theories were but lightly considered, but the editors urged enterprising Americans to continue the search. "The matter is of too thoroughly vital importance to be dismissed simply because difficulties are in the way," they urged. "Store the Wind Power."[30]

Suggestions and theories kept on coming. A reader suggested that wind wheels be used to drive dynamo-electric machines to decompose water into hydrogen and oxygen, the constituent gases to be stored in containers for use in heating and lighting.[31] One miner (as recounted in chapter 1) offered the example of a sand-storage system to power his arrastres.

Another reader came up with an idea still used today; namely, using the wind machine to raise water to a reservoir, then releasing the water as needed for power production. The reader correctly noted that with a water reservoir "you are dealing with a commodity which can be better controlled than a heavy system of weights."[32]

The editors of *Scientific American* did not come to a firm conclusion; that is, they did not identify any suggestion as offering a practical, efficient method of storing the energy of the wind. As journalists of technology, they hoped to sow seeds that could be reaped in due time. Although their hope was not to be fulfilled, to their credit they recognized the wastefulness of allowing a diffuse but abundant energy source to go virtually untapped. Their curiosity may have been stimulated by conservation concerns, particularly the possibility of exhaustion of the nation's coal supply. For the first time in modern North American history, a few leaders and professional planners questioned "the myth of superabundance." Waste, so much a part of nineteenth-century American development, they maintained, must be replaced by wise use.

More likely, however, the editors' interest sprang from their inquisitive nature. The late nineteenth century has been characterized as a time of "technological torrent." "No other nation," claims historian Thomas P. Hughes, "has displayed such inventive power and produced such brilliantly original inventors as the United States during the half-century beginning around 1870."[33] Much the inventor's effort centered on progress in the use of electricity. Perhaps the most profound influence on everyday life was the emergence of this new power. An unknown mystery at the time of the Civil War, electricity, by 1920, was known to all Americans. Most of them were dependent upon it. The debate on wind-energy implementation was part of this revolution. Here was a problem and they sought a solution.

BRUSH'S MAMMOTH DYNAMO

The wind dynamo was built to go for twenty years and it never failed to keep the batteries charged until I took the sails down in 1908.

—CHARLES BRUSH, 1929

One prosperous inventor and scientist who may have followed the exchange of ideas with great interest was Charles Brush—we have no way of knowing whether or not he did in fact read the *Scientific American*. In any event, while others wrote letters and debated, he designed and built. Historians of science credit Brush with several significant inventions, but ironically, in the long view of history, his most bizarre invention—his great wind dynamo—may be his most lasting legacy. Observing today's sixteen thousand wind generators spinning on the California hills, one cannot but recall Charles Brush—the first person in the world to use the natural power of the wind to create electricity.

Brush built his mammoth wind generator over the years 1887 and 1888 in the middle of his five-acre backyard. Neighbors, promenading along Cleveland's fashionable Euclid Avenue, might have been dumbfounded had they looked carefully behind his palatial home to see the wind turbine efficiently spinning. Windmills were common enough, but this one did not grind grain or pump water: it created electricity. In those days, electricity was mystery enough, let alone electricity created from the wind. Almost certainly, some who viewed the remarkable sight must have thought its creator somewhat daft: a man of wealth who had failed to outgrow his youth.

Quixotic though the idea might have been, Charles Brush was no Rube Goldberg inventor with a head full of dreams rather than reality. A tall, dignified, mustached man, he had already made a fortune with his arc-light system and his numerous patented inventions—a formidable scientific legacy.[34] Since boyhood, he dreamed of capturing the indirect energy of the sun[35] and to do so he built a wind-electric generator—one of such scale that it provided power for 350 incandescent lights, two arc lights, and a number of electric motors.

With plenty of engineering savvy and a good deal of cash, Brush assembled a sixty-foot tower, weighing 80,000 pounds, and put together a complicated system of gears and pulleys so designed that the dynamo made fifty revolutions to one of the wind wheel. His basement housed the storage setup: a complex of twelve batteries, each with thirty-four cells. The system—a complex of gears, girders, belts, and instruments—was extraordinary, the first of its kind. And

Brush's massive wind dynamo dominated his five-acre backyard. To gain a sense of the machine's size, note the garden worker cutting the lawn to the right of the windmill, with the Brush mansion, also to the right of the windmill, in the background. The large, rectangular form behind the wheel, to the left, is the tailvane, which helped the tower to rotate with the wind. (Courtesy Western Reserve Historical Society, Cleveland, Ohio)

it worked. "As an example of thoroughgoing work it cannot be excelled," the editor of the *Scientific American* rhapsodized.[36]

That Brush could design and construct such a successful machine would not have surprised those who knew his background. Few Americans were better prepared to make contributions in science and technology. Born on March 17, 1849, on the Brush family's Walnut Hills Farm near Euclid, Ohio, as a youth Charles preferred his workshop to the surrounding woods. While other young men spent their leisure time hunting, fishing, riding horseback, or courting young ladies, Brush was honing a natural turn of mind toward science and invention. With his father's encouragement, he collected odds and ends from around the farm, reforming them into various mechanical gadgets.[37] His most successful creation was a velocipede—a sort of tricycle, constructed from two wheels taken from a baby buggy and one from a threshing machine. At age eleven, he and his machine were often seen keeping pace with the local horse and buggies. Such contrivances attracted attention, but there were also developing in the budding scientist's psyche elements that would make him unafraid of facing intellectual challenges. Later in life, he recalled that as a youth "I was an omnivorous reader of scientific literature. Such parts of astronomy, chemistry, and physics as I could understand were a never-ending delight." Early on, he applied this knowledge, constructing, as he put it, "much crude apparatus—telescopes, microscopes, and photographic appliances."[38]

In today's parlance, Brush was a gifted child. His years in high school were a time of experimentation with electrical apparatus, such as static machines, Leyden jars, batteries, electromagnets, induction coils, the arc light, and small motors.[39] As a senior at Cleveland High he was responsible for all the school's laboratory apparatus. When he graduated, he gave an oration on "The Conservation of Force"—tracing the course of solar energy through vegetation, to coal, to steam, to electric current, and finally to light.

His speech—which no doubt impressed the parents and perhaps baffled (and bored?) his classmates—set a major theme in his life.[40] The subsequent building of the wind dynamo must be understood in the context of seminal ideas expressed at that high-school graduation in 1865.[41] Brush went on to matriculate at the University of Michigan, earning his degree at an accelerated pace. Drawn as he was, almost magnetically, to electrical engineering, he found that such a curriculum was not offered so he did the best he could, majoring in mining engineering.[42]

Following his formal education, Brush dabbled in an unprofitable chemical laboratory, then soon turned to the iron-ore and pig-iron business for a livelihood. Uncreative work, however, was less than satisfying for Brush, who passionately wanted to invent. Soon he formed a partnership arrangement with George Stockley, of the Telegraph Supply Company of Cleveland. Stockley provided the space and equipment and Brush worked in his spare time. This collaboration bore fruit when Brush developed a more advanced electric dynamo. Soon he combined his Brush dynamo with his most significant invention, the Brush arc light. A public demonstration in downtown Cleveland of the new lighting system created a sensation, and the news reverberated throughout the country. Municipalities and private enterprises such as factories requested installation. By the close of 1880, some six thousand Brush arc lights were in operation throughout the nation, bringing illumination to previously ill-lit cities.[43] The Brush Electric Company (a forerunner of General Electric) was successfully launched, and Brush's fortune was assured.[44]

In only a few years, the talented inventor rose from his relatively modest background to esteemed status. Wealth poured upon him like gold upon a lucky argonaut and he designed and built the Euclid Street mansion mentioned earlier, taking his place among Cleveland's wealthy. For some, such sudden wealth might have spelled the demise of work and creativity, but for Brush fortune meant freedom: freedom to pursue science and invention without his projects having to have marketable, practical application. Although Brush's financial status changed, his work habits did not. He spent long hours in his basement laboratory, running a variety of experiments, and transforming the successful ones into bold ventures. The date of conception of the plan for a giant wind generator is unknown, but in an unpublished biography of Brush, Margaret Emily Richardson wrote that, once his house was completed, he could turn his attention to "a Don Quixote dream of an enormous windmill to light his house and turn its various motors . . . an idea which had fascinated him from boyhood."[45] When Brush died, in 1929, one of the eulogies offered gave the colossal windmill a more significant relationship to Brush's life. The wind generator, said the orator, "symbolizes . . . the two primal forces motivating him from within. First, the sun-power vision expressed in his high school oration, and second, the desire to accomplish this at the minimum of human effort. The old windmill came from the veritable shining center of him."[46]The fact is that we lack written record that might give Brush's view of his motives. In his published works the wind generator is not mentioned. There is also no evidence as to motivation in Brush's personal papers.[47]

In the absence of authoritative explanations, it will be worth while to plunge into calculated speculation. Surely, Brush's fascination with natural forces, and particularly those emanating from the sun, influenced his decision. His practical streak played a role. Might not a useful, marketable wind generator have been designed from such a prototype? Brush's personal need for electrical power, however, furnishes the most logical explanation. In the mid-1880s, central power systems did not exist—not in Cleveland or any other U.S. city, save a small area of New York. In fact, nationwide, only a few well-heeled people could boast a home lighted by electricity.[48] Thus, Brush may have been motivated by vainglory; but it seems

Charles F. Brush, of Cleveland, made a fortune through his Brush electric light system. Later, in 1886, he constructed the first windmill designed to produce electricity. (Courtesy the Western Reserve Historical Society, Cleveland, Ohio)

more likely that the electrical requirements of his basement laboratory launched his inventive mind in the direction of wind energy.

Not only do we not have record of Brush's motivations; also lost to history are documents explaining the wind generator's planning and construction phases. Presumably, such documents existed but have, over time, been misplaced.[49] What a fascinating undertaking the early stages of the dynamo must have been! And what a sensation Brush must have felt when the mighty wind machine first began to turn, transforming vagrant wind into useful electricity.

Unfortunately, we are limited to knowing that in 1888 the wind generator began operating, and that to electricity buffs, it was state of the art. In terms of wind-generated electricity, nothing comparable had been attempted in the United States. As noted earlier, the idea of wind-generated electricity had been a subject for discussion in such magazines as the *Scientific American*, but no one before had translated the theoretical into the practical, and certainly not on the scale of the Brush wind generator, with its sixty-foot high tower, and a wind rotor fifty-six feet in diameter, with 144 wooden blades, and 1,800 square feet of total blade-surface area.[50] Not only was the scale of grand proportions, but the technology was striking for its day. As noted earlier, Brush had designed an elaborate gear-and-pulley system, allowing the dynamo to make fifty revolutions for every one of the wind wheel or rotor. Thus, he compensated for a slow-turning rotor (often a problem in the generation of electricity) by using an advanced gear-and-pulley system that allowed a 50:1 gear ratio.

The dynamo, or electric generator, specifically designed by Brush, would produce 12,000 watts of direct current (DC) electrical power at a full load of five hundred revolutions per minute. The switching devices and safeguards were particularly innovative. The generator kicked into production at 330 revolutions a minute, and an automatic regulator prevented the electric power from running above 90 volts at any speed. The working circuit was set to close automatically at 75 volts and open at 70 volts. Brush built other

SCIENTIFIC AMERICAN

A WEEKLY JOURNAL OF PRACTICAL INFORMATION, ART, SCIENCE, MECHANICS, CHEMISTRY, AND MANUFACTURES.

NEW YORK, DECEMBER 20, 1890.

THE WINDMILL DYNAMO AND ELECTRIC LIGHT PLANT OF MR. CHARLES F. BRUSH, CLEVELAND, O.

The *Scientific American* journal devoted the cover of its December 20, 1890, issue to Brush's wind dynamo. A lengthy article about the sixty-foot, forty-ton machine said admiringly: "As an example of thoroughgoing work it cannot be excelled."

less obvious safeguards into the electrical system as well as the mechanical apparatus. A *Scientific American* reporter admired the fact that "every contingency is provided for, and the apparatus, from the huge wheel down to the current regulator, is entirely automatic."[51]

Reporting on Brush's feat, the *Scientific American* failed to mention a historic connection—Brush's wind generator followed the general principle of the English post-windmill perfected in the thirteenth century. A sturdy, fourteen-inch wrought-iron gudgeon or post was sunk eight feet into solid masonry. On this post sat the iron frame of the 80,000 pound tower. To support the tower during heavy wind, arms pointed downward and outward at each corner to a circular track. At the end of the arm was affixed a caster that, under normal circumstance, came close but did not touch the track. However, when the wind velocity quickened, the caster would make contact with the rail, relieving the center post of further pressure and possible damage.[52]

For this reliable source of wind-generated electricity, Brush required a storage system and, on this subject, Brush was himself one of the most knowledgeable scientists in the United States. In the 1870s, he had pioneered electrochemical research, and that work had resulted in significant, and sometimes controversial, patents.[53] To store power, he installed in his basement a bank of twelve batteries, each consisting of 34 cells, or a total of 408 cells. From this storage system he commonly used about one hundred incandescent lights of sixteen- to twenty-candle power. More demanding on the storage system were two Brush arc lights and three electric motors.

The wind-dynamo electrical system worked remarkably well. Brush quickly mastered a technology that even in the present age is fraught with difficulties and frequent failure. According to the *Scientific American* correspondent, when he visited Brush the system had been in "constant operation" for two years.[54] Although there are no records extant, we know that the wind dynamo operated throughout the 1890s. At the turn of the century, Brush voluntarily abandoned his private wind-power plant for the convenience of using centrally produced electricity. Whether he had doubts or agonized over this decision is a matter for conjecture.

Certain questions arise regarding Brush's creation that can never fully be answered, but nevertheless deserve attention. First, we might ask why Brush never entered a patent on the wind dynamo? Brush patented numerous inventions and apparatus, and as an inventor he understood the importance of protecting his creations. We can only suspect that Brush realized that his wind dynamo had no commercial value, at least not in Cleveland, where the intermittent wind necessitated a huge and costly bank of storage batteries. A friend and scientific colleague, Charles Baldwin Sawyer, later wrote that "the windmill-generator, with power reservoir of a storage battery might, he hoped, be useful throughout the world."[55] Surely, Brush must have considered the value of the idea, but he understood that his wind dynamo could not be easily reproduced. It was on such a grand scale as to be affordable only to the wealthy. The *Scientific American* was correct in warning its readers that they "must not suppose that electric lighting by means of power supplied in this way is cheap because the wind costs nothing. On the contrary, the cost of the plant is so great as to more than offset the cheapness of the motive power."[56] Not only was it costly, but the machinery required knowledgeable care.

Another significant factor centered on the emergence of cheap, reliable, mass-produced power. Urban residents, such as those in Cleveland, would soon be served through central power systems.[57] The Brush windmill did not lend itself well to the burgeoning electrical rage. If anything, it represented individual enterprise, self-sufficiency, and the simplicity of exploitation of natural forces. The nation was taking seven-league steps in a different direction; namely, centralized power production and highly technological systems. In many ways, the Brush windmill represented a nineteenth-century concept that could not fit into twentieth-century complexity and interdependence.[58] No doubt Brush concluded that his wind experiment proved challenging and novel, but it could not result in any profitable enterprise—at least not in his lifetime. It would have been foolhardy, indeed, to risk his considerable fortune on such an unexplored and demanding technology. It was satisfaction enough, as the *Scientific American* reporter put it, to make "use of one of nature's most unruly motive agents."[59]

Although Brush was well-known in the forest city, national acclaim on the scale accorded to Thomas Edison and Alexander Graham Bell eluded him. Perhaps overshadowed by Edison, Brush was something of a scientific recluse. He was modest regarding his accomplishments and he often shunned publicity. He maintained an office at the Cleveland Arcade, but according to one account the sign on his door announced, "Office Hours, 11:30 to 11:45."—a hint to callers to make their visit brief.[60] Such brevity might be seen as due to efficiency; others might see it as impatience and irascibility.

Whatever his other personal characteristics, we know that for Brush work was an addiction, and experimental science was his narcotic. He was rather secretive about his experiments, toiling alone without assistants. He certainly shared his findings, but he perceived his scientific work in terms of individual creativity. Unlike contemporary inventors such as Elmer Sperry or Edison, who profited from knowledgeable and creative assistants, Brush refused to hire anyone. Once, when urged to employ an assistant, he rejoined, "Would an artist hire an assistant to paint his pictures?"[61] Such an answer represented a romantic sentiment, but not one out of keeping with his idea of the independence of the inventor. Into the twentieth century, he continued work in his basement laboratory, experimenting with gases and gravitation. After dark, passersby would have seen lights ablaze behind his basement windows, for

Brush often worked in the late evening, claiming that daytime streetcar vibrations flawed his apparatus.[62]

Pride in his windmill—which had become a Cleveland landmark—was evident when he informed a Cleveland reporter in the late 1920s that it "was built to go for twenty years and it never failed to keep the batteries charged until I took the sails down in 1908." The seasoned inventor went on to say that although he hooked up to city current in 1900, he continued to use the windmill-charged batteries for his laboratory.[63]

Late in life, Brush broadened his interests. Like many other men of wealth, he invested in real estate, being particularly drawn to the Cleveland Arcade, a five-story glass building, the length of a city block—a forerunner of the modern shopping mall.[64] He attended lectures, served on the board of trustees at Western Reserve University, delivered scientific papers, and belonged to three prestigious Cleveland clubs. He also contributed to many philanthropic causes.[65] A robust, six-foot-two, broad-shouldered man, he cut a handsome figure as he made his way about the streets of the city he loved.

When Brush died on June 15, 1929, all that remained of the windmill was the sixty-foot-tower, worse for wear, standing somewhat decrepitly in the spacious rear yard. Although Brush commanded a comfortable, indeed enormous, income, he chose not to maintain his technological wonder. He had dismantled the wind wheel, stored the dynamo, and allowed the belts to rot[66]—a sad fate for a curious machine that had become a Cleveland conversation piece.

The forlorn windmill became the source of some controversy. Although Brush had bequeathed all of his scientific and laboratory apparatus to Charles Baldwin Sawyer, he made no specific mention of the wind dynamo in his will. All agreed that Sawyer had the power to dispose of the items entrusted to his care; however, special circumstances were revealed about the wind dynamo.

On January 25, 1929, James W. Bishop, of Dearborn, Michigan, had called on Brush. This well-heeled visitor, representing Henry Ford's antiquarian interests, proposed that he and his staff move the Brush windmill to the Museum of American Antiquities and the Edison Institute for Technology, both in Dearborn. It would be restored and maintained in operating condition. According to the sworn affidavits of Brush's secretary and another observer, Brush agreed to the request.[67] Brush had no doubt assumed that the old tower, relatively ignored for so many years, would be torn down: he had made provision in his will that his mansion would be destroyed if no member of his immediate family wished to live there—and why should the windmill be spared? When Brush promised the deteriorated tower to Bishop, he was surely flattered that a respectable museum director wanted it.[68] In all likelihood, the remaining structure and the surviving apparatus would have been shipped off to Dearborn. However, Henry Ford was out of the country and the busy automobile tycoon simply never got around to authorizing the acquisition of the Brush windmill. Thus, no written acceptance of the offer was made. Little more than four months later, Brush died.

Often we fail to realize the value of a possession until someone else wants it—an idiosyncratic behavior that now possessed Cleveland. The windmill became an artifact not only associated with Brush but with civic pride. The Cleveland *Plain-Dealer* headlined a story about the controversy, "City and Ford In Clash Over Brush Relic."[69] Councilman Ernest J. Bohn insisted that the city keep the windmill and place it in a public park because of its historical interest. Others noted that Ford's Dearborn museum would primarily honor Thomas Edison, Brush's competitor. Executor Sawyer expressed the concern of many of Brush's admirers when he wrote that "it does seem a pity that the windmill or any of this material should go to Dearborn, if it is to be used principally to add to the glory of Mr. Edison."[70] Later, Brush's biographer was less diplomatic, charging that Henry Ford's attempt made "Mr. Brush seem like the inventor of windmills instead of the first practical electric light."[71] Such a conspiratorial theory is based more on suspicion than fact.

Throughout the summer of 1930, advocates for both Dearborn and Cleveland exchanged letters. James Bishop concluded a letter to Dr. Roger Perkins, a Cleveland associate of Brush's, with thinly-veiled irritation: "Don't you honestly think that his [Brush's] wishes should be carried out?"[72] Perkins responded that although he appreciated "the value of the Dearborn museum . . . where it is possible to retain a memorial among the natural associations, this is preferable."[73]

Sawyer, Perkins, and the city of Cleveland won the debate. In November 1930, Bishop capitulated, writing that "Mr. Ford decided that he would rather let the citizens of Cleveland have the windmill than to put up a fight for it."[74] He hoped that there would be no difficulty in finding the money to restore the old mill. Here, of course, Bishop raised the age-old problem of historical restoration—it costs money. The Case School of Applied Science (now Case Western Reserve University) was willing to provide space for the relic, but that was all. The president declared the university could not "meet any of the expense of moving, restoring, or installing this equipment." Nor could it pay for maintenance.[75] Given the space, the Cleveland Chamber of Commerce joined the preservation effort and interested parties formed the Brush Memorial Fund. The founders hoped to secure subscriptions "of about $100,000 . . . to provide for the moving, erection, reconditioning and upkeep of the scientific apparatus and the windmill."[76]

Documentation of the details of the Brush Memorial Fund effort has become muddled. What is clear is that, with the stock market crash of October 1929 and the nation moving toward the Great Depression, money to preserve the windmill did not come easily. Just how much was raised is uncertain. One letter from Alexander

Brown, president of the Brush Memorial Committee, to O. P. Van Sweringen summarized a meeting held on July 3, 1930. Apparently, Van Sweringen and his brother verbally pledged $60,000 to see that the windmill and equipment remained in Cleveland.[77] Whatever the outcome of that pledge, it did not result in action. Through 1930 and 1931, both weather and urban development threatened the tower and by November 1931, Alexander Brown was expressing appreciation to the president of the Ohio Bell Telephone Company for its offer to take down the tower, box it, and store it. The intention to reconstruct the windmill had not been altogether abandoned, but in 1931 dismantling and storing was the best available choice.[78]

In the years that followed, the commitment to restore Brush's inventions waned. Money was scarce. Interest in wind energy had diminished as centralized power systems replaced the rural wind generators. The old Brush windmill, under Sawyer's care, remained crated and sheltered behind the Brush laboratories on the mansion site.[79] It was never to be reconstructed. In the late 1950s, with Sawyer away, a zealous Clevite Company executive, looking for storage space, had the windmill tower removed and destroyed.[80] Had Bishop and Henry Ford prevailed, the Brush windmill might still stand. Cleveland and the Brush Memorial Fund won the legal battle but lost the windmill.

Brush was surely one of America's preeminent, patrician inventors—a man who turned his genius to practical account. He understood what could be marketed and what ought to be considered a creative novelty. In his earlier days, the great wind generator usefully met his electrical power requirements. But Brush fully understood that in a technological world of huge electrical systems there was no place for such an independent system. From a later perspective, however, this seminal machine can be seen as a precursor of the future. Although the technology that Brush employed bears little resemblance to today's path breaking wind-energy plants in California, the new and old are joined by their purpose: to create energy from a natural, free, nonpolluting, renewable resource.

The Brush windmill proved the basic feasibility of a marriage of wind and technology to produce electricity. Seminal machines, such as his, are seldom useful to the public, but they provide the inspiration and the challenge for the next wave of inventors.

CHAPTER 3

EXPLORING THE POSSIBILITIES

Electricity is "a dangerous servant liable to strike its master dead at a single unguarded touch."

—*RURAL NEW-YORKER*, June 9, 1888

THE DECADES BETWEEN 1890 AND 1920 REPRESENT THE PAUSE BETWEEN invention and application. The Brush wind machine could demonstrate that the kinetic energy of the wind could produce electricity. It was practicable, but was it practical? Could wind power become a part of the electric-energy mix? And could it, for isolated consumers, serve as the primary source? The answer to both questions was no.

At the turn of the century, wind energy looked like a whole brier patch of possibilities, but unfortunately the brier element triumphed over the possibilities. All was not lost, however, and wind-power advocates made some seminal, halting steps toward recognition. Engineers and industrialists recognized the many disadvantages of wind energy, especially in an energy world that demanded unlimited supply without consideration of environmental effects. In that era of abundance and waste they determined that there was no niche for such a marginal, undependable, diffuse energy source. Wind energy simply could not meet the technological criteria for use in factories or urban areas.

The technology did, however, appeal to creative, practical minds because it offered—as was often mentioned—an ostensibly gratis energy source, compliments of nature. The dictum Waste not, want not was embedded in the American character. Furthermore, a few inventors recognized the wisdom of capturing natural energy sources. Americans courted technology, but they did not divorce nature. The objective of each American, exhorted the editor of *Scientific American*, ought to be "to control the natural forces by which he is surrounded and then turn to useful account by making them the ministers of his comfort, profit, and pleasure." In an earlier age, these natural forces—wind and waves, mighty rivers and storm clouds—were viewed with superstition and awe. In twentieth-century America, they were under "cheerful submission" to human needs and desires.[1]

Cultural forces were also at work. In all eras, segments of the population question the direction of the nation, and Americans of the so-called progressive age were no exception. Many felt that the nation was beset with crises and paradoxes. New millionaires and a burgeoning upper class enjoyed their wealth; the less fortunate wallowed in poverty. Although the North had triumphed in the Civil War, blacks had been virtually reenslaved in the South. From Chicago to San Francisco, empty spaces were filling up with towns and land-hungry people. Cities exploded with growth as the nation buzzed with industrial maturity. A people whose forefathers had put down roots on farms now moved to the city. Those who had gauged their lives by nature's seasons and the rising and setting of the sun now found life metered by the repetitiveness and regularity of the time clock and factory whistle. Of course, there were many attractions to the new urban life, at least if you were white, male, and middle or upper class.

As might be expected, some Americans questioned these new directions, eschewing the city, industrialism, and materialism. Cultural rejection took different forms, but the skeptics shared the impulse to question. One cadre embraced the back-to-nature movement.[2] Many back-to-the-land advocates followed prophets such as Liberty Hyde Bailey, a sectarian preacher of the joys and benefits of the unadorned rural life, and for Bailey followers, intermediate technology provided assistance without being offensive to their lifestyle. The internal combustion engine and centralized electrical systems were suspect to those who questioned the accelerated pace of technology, whereas a wind machine might occupy some middle ground. As intermediate technology, it was understandable on a personal scale. Like the ideal of rural life itself, wind units were detached from urban influence and central systems. They helped minimize reliance on an increasingly complex and dubious society. In truth, many rural people were content to do without electricity, but for those who wished the benefits of power without the vulnerability of dependence on a central system, generating one's own electricity provided some measure of a solution.

While American capitalists committed to central power systems, European nations outdistanced the United States in the wind-energy field. Researchers in Denmark, Germany, and Britain all actively pursued wind-energy experimentation. The foremost world expert was not Charles Brush but Professor Poul LeCour, a Dane whose experimental work was occasionally covered in U.S. scientific and engineering journals. With the financial backing of the Danish government, LeCour both experimented and built. By New Year's 1906, some forty windmills were generating electricity in the small nation of Denmark, the beginning of a tradition that would last to this day.[3]

To the south, across the Schleswig-Holstein border, German engineers capitalized on LeCour's research as well as innovations introduced by an engineer named Hermann Honnef. In the town of Husum, engineers constructed a fifteen-horsepower (15 hp) unit that provided direct power to turn machinery; it was not for producing electricity. Near Hamburg, however, a forty-foot rotor designed to provide electricity to power a small plant operated so efficiently that for a period of ninety days the owners found it unnecessary to use the gasoline-engine backup system.[4] In 1905, German engineers continued their experimentation with a 50 hp wind-electric system—a device that proved to be the crowning achievement of early German wind engineering, providing electric lighting and general power service for the small community of Wittkeil in Schleswig. The editor of the *Scientific American* noted: "In Germany, the most economical of all powers [the wind] has been developed to a point that is surprising."[5]

Across the North Sea the British were not far behind. In the 1890s, the Rollason Wind Motor Company constructed wind machines with 5 hp dynamos for use in the English countryside. The rotors had only five blades, a system clearly superior to those of multi-bladed machines. Professor R. A. Fessenden, the English engineer with creative and somewhat futuristic ideas who led the project, proposed to build a series of large windmills on coastal cliffs. These behemoths would lift seawater for storage in reservoirs, whence it could be released to turn turbines and dynamos. Such a system, he prophesied, would produce power sufficient "to drive all the manufactories, railroad trains, and motors of all kinds in Great Britain and Ireland at considerably less cost than this was done by steam."[6] Certainly, this was a pipe dream, and yet such innovative dreaming (as well as practical application) was undoubtedly more forward-looking than what was happening in the United States. A few individuals recognized the fact. One reporter, lamenting that the United States had been left in the technological dust, noted that in Europe wind generators were a source of creative thought and practical use, while in his country the "scope of duty seems to be confined almost entirely to . . . water-raising devices."[7]

Of course, there was an explanation. American water-pumping windmills represented a proven and profitable technology—an American success story. By the close of the nineteenth century,

U.S. manufacturers had produced over a million windmills. Known collectively as American windmills, these machines could be found on almost every Western farm as well as in all corners of the globe.[8] Lightweight, self-oiling, movable, and inexpensive, American windmills dominated the world market and had become one of the nation's notable export products. Wind-electric generating systems, on the other hand, were experimental, unreliable, and something of an enigma to most consumers, and the technological mysteries were daunting to all but the most daring innovative mind.

American inventors were not completely lacking in imagination, however. One correspondent to *Scientific American*, in 1892, predicted the future direction of wind-energy development. Noting the "incessant" nature of the wind in the American West, he challenged readers to "think what an enormous force could be created by some twenty large windmills cooperating."[9] Here was a seminal idea. It was the first reference to the use of wind turbines working in unison as a wind farm or generating station. But in the 1890s, such a notion seemed ludicrous. No scientist had performed for the United States the service that Professor LeCour had done for Denmark; that is, to determine suitable (i.e., windy) harvesting sites. Geographers had of course recognized North America's windy areas—the coastal regions, the mountains, and the West—but wind-farm development required systematic study. In addition, scientists had to develop a more efficient, less expensive storage battery, otherwise wind machines could only serve in tandem with another, more reliable generating source.

There was also another barrier to development. From the time Thomas Edison built New York's Pearl Street Station in 1882, engineers understood that city electrical service would be provided through centralized systems. Farmers and ranchers, being outside such systems, represented the potential market for individual wind-energy units—but many farmers were reluctant to embrace change, especially with regard to electricity. Historian Clark Spence explained that U.S. farmers "were innately conservative and skeptical of new devices." The mysterious force called electricity evoked fear rather than longing. Many rural residents must have nodded their heads in agreement with the *Rural New-Yorker* editor who warned his readers that electricity "is a dangerous servant liable to strike its master dead at a single unguarded touch."[10]

Although cost and the state of technology prevented the electrification of many rural areas—to use a word that came into vogue—the cultural factor was of equal significance. In a percentage of the rural population, change, no matter what the advantages, would run into resistance. In 1913, a committee report to the National Electric Light Association lamented that "practically nothing has been accomplished in real extension of electric lines to the rural districts," citing the "lack of faith" of rural consumers as one reason for this.[11]

Some farmers, however, quickly overcame their reservations and incorporated electrical power into their operations. In regions that, in spite of being semiarid, sustained intensive agriculture (such as the Central Valley of California) farmers hailed electricity as a guarantor of prosperity. These yeoman growers needed water to irrigate their fields, and water-pumping windmills were inadequate; they needed electric pumps. These farmers were able to consume such quantities of electricity that the Pacific Gas and Electric Company found it profitable to string lines and give them the benefits of centrally produced power. Long before farmers elsewhere had that option, Central Valley cultivators had unlimited access to electricity. One result was that whatever incentive existed for individual wind-energy units died, and scarcely one such unit could be found in California.[12] Few, if any, cultivators opted for independent power if a central source was available.

In time, some farmers noted a loss in their sense of self-sufficiency and also questioned the high cost of company-provided electricity, but initially these factors seemed inconsequential compared with the benefits. Farmers, often presented as a bastion of individualism, self-reliance, and traditional values, seemed to ignore this time-honored commitment to independence. The view that small, decentralized systems might be better than huge operations seldom entered the debate.[13] But enter it did—and from an unlikely source. William Smythe, the great proponent of government-sponsored irrigation systems, admitted that with water supplies there were trade-offs. "Small individual pumping-plants have certain advantages over the canal systems which prevail elsewhere," he conceded. "The irrigator has no entangling alliances with companies or co-operative associations, and is able to manage the water supply without deferring to the convenience of others, or yielding obedience to rule and regulations essential to the orderly administration of systems which supply large numbers of consumers."[14] Of course, for Smythe the gains of centralization in water supply outweighed the losses; reliability was favored over shaky assurances, and bureaucracy was preferable to the mythology of individualism.

"THE OIL LAMP MUST GIVE WAY"

The future may see nature harnessed in new ways—
from the sun, wind, or sea—who knows?
But one thing is generally accepted:
that new methods will be combined with electricity.

—GENERAL ELECTRIC COMPANY AD, 1913

Understandably, wind-energy advocates were more or less impotent in the face of the arguments for central power. Engineers and scientists believed implicitly in the economies and efficiencies of scale. They also were committed to centralized systems, and—in technological terms—those systems may be a defining characteristic between nineteenth- and twentieth-century America. Technological individualism became anathema to the emerging world; complex systems became a sought after blessing—a benediction of scientific arrival.[15]

In spite of the obstacles, a few companies ventured into the business of producing small wind-energy units. These companies had experience in producing water-pumping windmills, and they viewed this new market, although small, as potentially significant. Like other entrepreneurs, they recognized that the nation would demand more and more electrical energy. Most, but not all, of that energy would be created by fossil fuels and hydroprojects (falling water). They might have taken heart from an advertisement put out by the newly created General Electric Company: "The future may see nature harnessed in new ways—from the sun, wind, and sea—who knows? But one thing is generally accepted: that new methods will be combined with electricity."[16]

The Lewis Electric Company of New York was the first to offer a commercial wind-energy conversion system. By November 1893, the company provided dynamos for use "in wind mill service for house lighting." It is difficult to assess the number of units installed or the success of the Lewis system,[17] but in 1894, a magazine, the *Electrical Engineer*, was impressed enough to feature one installed by Bostonian George E. McQuesten. In the spring of 1892, McQuesten had installed an 8 hp steam engine at his Marblehead, Massachusetts, summer home. It provided power to light his home, the grounds, and outbuildings. However, McQuesten had become dissatisfied, because his gardener had to spend so much time firing up the engine and generally looking after the system. Hence, in early May, 1894, the efficiency-minded owner replaced the steam engine with a twenty-foot Eclipse wind wheel, mounted on a seventy-five-foot tower. The windmill turned a 3kW Lewis dynamo, which in turn charged forty-six Bradbury-Stone storage batteries. The system, which was equipped with automatic regulators and other devices, provided plenty of electricity for lighting

In 1893, the Lewis Electric Company, of New York, offered the first commercial wind-energy system for the production of electricity. It was offered for sale to the wealthy as a pleasing, yet functional, lawn ornament. It followed the general engineering principles of the Brush dynamo. (From the *Electrical Engineer* 17 (January 31, 1894): 86)

and, in addition, when there was sufficient wind, would run electric motors in McQuesten's shop. The system worked well, and that

November, when McQuesten closed the place for the winter, he was pleased with the wind generator's initial performance.[18]

In marketing the innovation, the Lewis Electric Company played upon a romantic theme that in future years would be adopted by many other companies:

> The smoking, oil-dripping lamp, with offensive odor and always more or less dangerous, must give way. The gentle wind that "bloweth where it listeth" gives motion to the wheel surmounting a graceful and artistic tower that ornaments the lawn, and, by the use of The Lewis Independent System of Electric Lighting, generates a current of electricity that fills the place with a blaze of light and lays up in reserve enough for times when the wheel does not turn. What a triumph! Lighting your home with electricity by wind mill power![19]

The ad provides clues as to the Lewis Electric Company's market. First, it was limited to the New York and the New England coastal regions (Long Island, Marblehead, and so forth). Second, the company expected to sell these independent wind-electric systems to wealthy tinkerers—such as McQuesten—and the upper-middle class. The company assumed the prospective buyer would have a yard and a lawn—a suitable site for the wind system. Also, the value of the machine was presented as more than the mere generation of electricity. The graceful and, compared with steam, quiet system was an aesthetic addition to one's home (as the Brush wind generator had been to his palatial mansion; in fact, although mechanically different, in illustrations the Lewis machines resemble the Brush tower in their dedication to discriminating placement in the setting of country estate homes). The fact that, probably, few of these elegant machines were installed makes it difficult to declare them the first commercial application of an independent wind-energy system.

A better contender for the title of first might be Fairbanks, Morse and Company, manufacturer of the Eclipse windmill, a pioneer, wooden water-pumping machine. In the 1890s, the company's engineers modified their traditional, multibladed rotor by adding a small dynamo and storage unit.[20] The Aermotor Windmill Company, manufacturer of the most popular water-pumping windmill in the United States, also entered the market. For a number of years, the Aermotor company had been capturing a larger and larger share of the windmill market and by the twentieth century its simple, reliable, inexpensive, steel-bladed machine would be omnipresent in the arid West. La Verne Noyes, a founder of the company, experimented with wind energy and soon became convinced of its potential. He moved the company toward wind-electric machines. By 1895, the New York office of Aermotor featured a roof windmill that provided current for electric lights. The future seemed bright and the problems slight. A Chicago-based publication, *Farm Implement News*, predicted that the simplification of electric-lighting apparatus would make "practicable for the ordinary geared wind mill to provide the power necessary for lighting farm houses and farm buildings, all at a cost within the means of the ordinary well-to-do western farmer." And with an eye on the manufacturers, the editor noted: "For the wind mill industry the future promises richer fields to work and greater prosperity."[21]

The *Farm Implement* article was the first in a string of optimistic reports based more on faith than facts. La Verne Noyes, grasping the difficulty involved in adapting the traditional American water-pumping windmill to the generation of electricity, invested in research and development. It was not until the early 1920s that the company offered the Electric Aermotor to rural America.[22] Caution and research were not rewarded, however: the new model—retaining, as it did, the multibladed wheel—never approached the success of the company's water pumper. Other companies, being less patient, modified slow-turning windmill wheels, producing marginally efficient and usually disappointing machines. These producers seemed to understand marketing but not engineering.

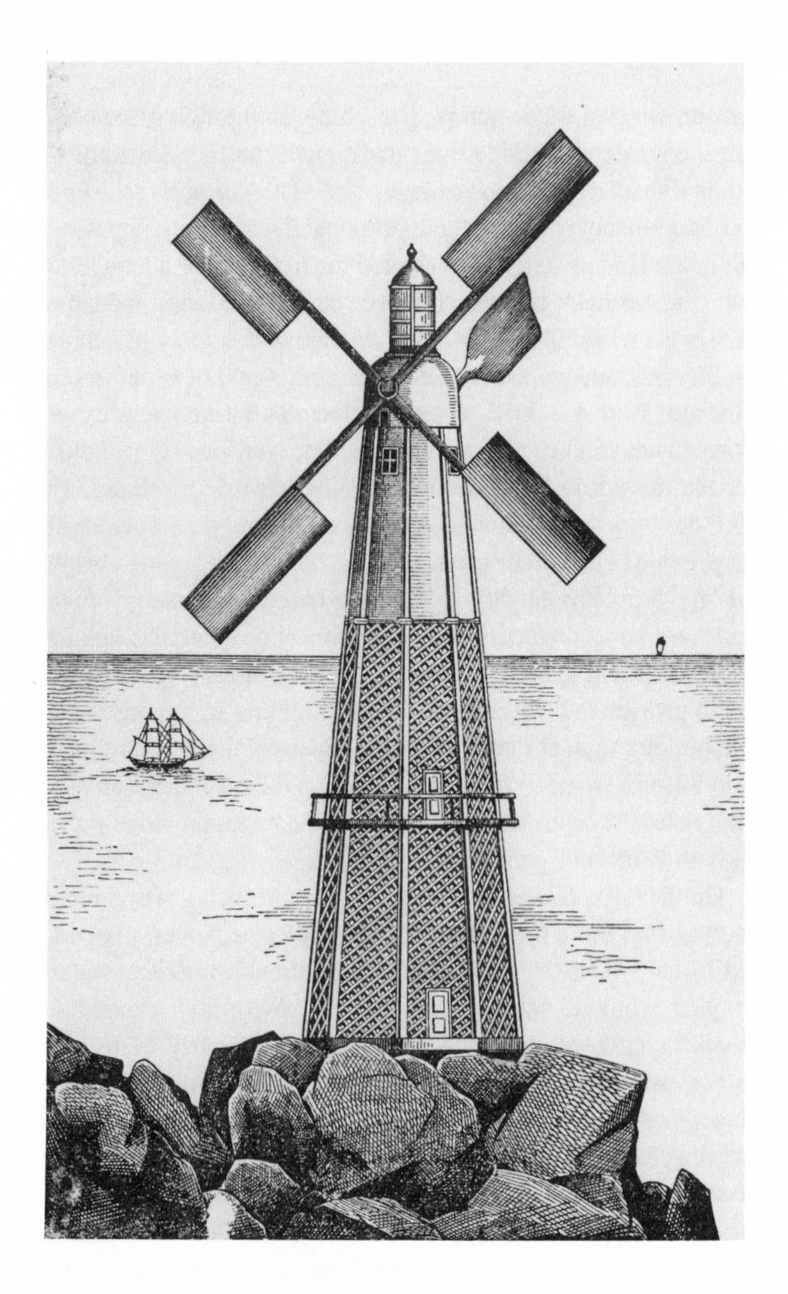

The structure shown in this 1890s sketch, "The Hannaford Electric Light House," was probably never constructed, but it indicates the perceived potential for wind energy in the late nineteenth century. (From the Challenge Wind Mill and Feed Mill Company trade catalog, circa 1895. Courtesy T. Lindsay Baker, Rio Vista, Texas)

The slow-turning wheel simply did not create enough revolutions per minute (rpms) to operate the generator effectively. These early, makeshift machines produced a legacy that the wind-energy business has not fully lived down even today.

In summary, U.S. engineers made little progress toward perfecting Charles Brush's wind machine. The Brush windmill became a relic of the past rather than a model for the future—for reasons that are quite understandable. Scientists and inventors viewed wind machines as a possible solution to *individual* energy needs, not those of society. Engineers looked to complicated systems of energy generation and distribution for electricity supply, both in the city and the countryside. Huge, interdependent electrical systems with equally complicated corporate structures provided the direction for U.S. energy development. The wind-farm ideas of the English engineer Fessenden found little audience in the United States, and even if they had, the gulf between theory and application was too great a chasm to bridge.

Even within the rural individual's electrical needs, a wind-energy unit was a questionable investment. Farmers found the machines unsophisticated and unreliable. A few farmers used them, and a smattering who were mechanically handy built their own. Some, insisting on having their own generating plant, used one powered by gasoline, kerosene, or fuel oil. Most rural people, however, simply waited—often in vain—for a hookup to a central system.[23]

For the wind-electric business, the fin de siècle was the first decade of the doldrums. The nation was moving in another direction. Central systems of energy production and distribution—"networks of power," as Thomas Hughes, historian of technology, called them—convincingly displaced any idea of meeting electricity demand through single-home or single-factory production. Huge utility cartels spread throughout the nation, and men like young Samuel Insull (whose entrepreneurial exploits are discussed in chapter 5) put together power leviathans that would mature in the heady days of the Roaring Twenties.[24]

It is ironic that during a period when Americans first became aware of waste and the need for conservation of resources, a technology that represented a partial solution was so ignored. In 1908, a U.S. president for the first time raised the specter of exhaustion of nonrenewable resources, such as minerals, iron ore, and coal.[25] Yet to expect American political and industrial leaders to have developed a difficult, diffuse, and marginal energy resource such as the wind would be to impose on that era sensitivities that did not then exist. In those days, whether it be tons of wheat, acres of land, kilowatts of power, numbers of skyscrapers, children in school, or miles of railroad lines, Americans spoke in superlatives that no nation in the world could equal. The cost to the environment of such progress was hardly a consideration.

CHAPTER 4

WIND ENERGY FINDS A PLACE

The AEROLECTRIC is as good as a maid for your wife—
it will do the washing—make ironing a pleasure, make cleaning a joy—
she can use a toaster, a percolator, curling iron,
and any of the many conveniences, which are operated by electricity.
No need to use it sparingly, because it is free—
it's only the wind in another form.

—PERKINS CORP. ADVERTISEMENT, CIRCA 1925

WAR IS ONE OF THOSE PROFOUND PARADOXES. IT IS DESTRUCTIVE TO human values, to life, to cultures, to natural resources. Yet it is also a time of technological and scientific progress. Amid the chaos and bloodshed emerge the inventions that become the tools of progress in the peace that follows. This was the case with war-accelerated research and the development of wind-electric plants.

The science of aeronautics proved to be the catalyst that contributed the most to this progress. The development of the airplane, spurred on by the exigencies of war, provided technological benefits that assisted a floundering wind-energy business. Specifically, increased understanding of air foils and propellers launched the industry in a very different direction.

Two mechanical-engineer scientists were most responsible: E. N. Fales and H. R. Stuart. Fales and Stuart carried out experiments to determine the windmill "best suited to radio-electric-generator drive, with particular reference to speed regulations."[1] Their windmills were small, even tiny. They were developed to provide aircraft with enough electricity to run a radio and, often, in addition, a fuel pump. "These little windmills," wrote one engineer, "had to aim at aerodynamic and mechanical perfection for their power came indirectly from the main engines and also weight was vital."[2]

Fales and Stuart worked under the auspices of the United States government aircrft program in 1917, at a Dayton, Ohio, laboratory.[3] Their tests "showed valuable possibilities of improvement in commercial windmills, for the aircraft windmills reached higher efficiencies than commercial practice had led one to expect."[4] The primary discovery made by Fales and Stuart centered on the rotor. The traditional multibladed windmill on farms had torque problems. Substitution of an airplane propeller-type blade, a product of aeronautic research, provided a revolutionary advance and mitigated these problems. "The new propeller," explained Fales, "differs from the others in its blades, which are one to four in number, of streamline cross-section, like that of an airplane propeller. The speed of revolution is from 6 to 10 times that of an American-type wheel of equal diameter in equal wind velocity."[5] The ensuing increased speed of revolution allowed for an efficient design that eliminated elaborate—and expensive—gearing systems: the rotor could be keyed directly on the armature shaft.

After the war, Fales continued his work with small-scale generators, constructing a wind-electric plant at a Ohio farmhouse. In spite of the unreliable Ohio winds, Fales reported that, because of its improved lightness, cheapness, and higher rotor speed, "competition with gasoline farm-lighting plants is now possible."[6]

The potential was enormous. Could such machines capture a share of the market? Electricity had evolved from a plaything of the wealthy to a necessity for the majority of Americans. Electricity seemed a miracle. The popularizer Waldemar Kaempffert compared electricity to Aladdin's lamp: "Aladdin rubbed his lamp and all things were possible of accomplishment. Today we press a button to achieve similar wonders. . . . The lamp and its owners were unreal. The electric spark is omnipotent, its powers everlasting."[7] Kaempffert looked to central power systems to perform such miracles, but he also lamented that engineers had spent so little effort on water power or wind power. Searching for an explanation for such neglect, he settled on the idea that "no really first-class inventor has ever associated his name with modern adaptations of the very ancient devices that depend on breezes or falling water."[8] Kaempffert's explanation may have validity. No prominent U.S. inventor, save Charles Brush, devoted much effort to the perfection of wind energy. And, as noted earlier, even Brush did not sustain his interest for long; he saw little utility in wind plants. Brush's research turned to compressed air and vacuum theory—areas he considered to be both more theoretical and more significant.

With the increased knowledge of the 1920s came a resurgence of interest in wind-generated electricity. As one writer put it, even though it often took a "good electrician to make the outfit work," there is "much appeal . . . in the idea of using something as free as the wind to generate power for farm lighting and light jobs."[9] Backyard tinkerers carried out experiments in barns, perhaps commandeering an old car battery, some copper wire, and an Eclipse or Aermotor windmill and fashioning them into a workable power unit.

Two young men from North Dakota, George and Wallace Manikowski, were typical. The two brothers toyed with windmill-electric generation for fifteen years. In 1920, they incorporated the Aerolite Wind Electric Company, producing a machine with a fourteen-foot wheel and the generating belt installed *around the outside of the rigid rotor*. The device also had regulators, generator, and storage batteries. All, according to the brothers, worked successfully. It was a novel design. However, the gearing system (which obtained forty-eight revolutions of the generator for each of the wind wheel) surely challenged their limited engineering ability. Supposedly, the Manikowski brothers' machine "proved commercially successful," but the evidence is scarce indeed, and the year following their incorporation (1922) they encouraged the Woodmanse Manufacturing Company to buy them out.[10] But wind energy continued to intrigue mechanically-competent farmers and others. A minister in Paoli, Colorado, enjoyed wind-generated electric lights in his parsonage and church.[11] Much of this activity occurred in the arid and semiarid West. Historians have often assumed that Western settlement was almost completed by 1890. Perhaps the conflict between whites and the Plains Indians had been reduced from violence to despair (and reservations) but much of the West remained unsettled. Vast regions remained free of human activity. Of course, population and financial capital did flow into some of the valleys of California and Utah, and with it water projects and electricity for irrigation. Over 43 percent of Utah farmers and 25 percent of California farmers had electricity by 1920.[12] But these enclaves were anomalies. Much of the West supported only a few hardy souls. They lived, one might say, in a state of technological deprivation. Certainly they had no central system electricity. Most of Montana, the Dakotas, Wyoming, Colorado, Idaho, Arizona, Nevada, western Texas, and New Mexico was livestock country, thinly populated, and ripe for wind energy.

Tinkerers in the West and elsewhere found encouragement through the first book published on wind-electric energy. In 1918, F. E. Powell's *Windmills and Wind Motors: How to Build and Run Them* offered information "in a branch of mechanics unfortunately much neglected."[13] The author encouraged his readers to build a wind motor and use it to provide electric lighting. The task would be simple and inexpensive, and the result would be reliable—although the author cautioned his readers that they "must be prepared to experiment a little," not with the principles of wind power but with the details of each individual apparatus.[14] To the publisher's credit, the book provided detailed descriptions of the apparatus, particularly the controls that would have to be installed. Understandable illustrations accompanied knowledgeable descriptions. The book appealed to the "model engineer" and without doubt an enterprising amateur electrician could follow the author's advice and build a serviceable plant.[15]

Meanwhile, commercial companies were providing other alternatives: mass-produced outfits featuring revised technology. Perhaps the first young entrepreneur to enter the field was Oliver Fritchle. This Denver-based inventor is best known for the Fritchle hundred-mile electric automobile, an electric car that would travel one hundred miles without recharging. He built five hundred vehicles between 1904 and 1917. Always fascinated with electricity, Fritchle had a particular interest in the potential of wind-electric plants. In fact, when he was a youngster, Fritchle's father took him to Cleveland to see Charles Brush's Euclid Street wind dynamo.[16] The visit excited his interest all the more. By 1917, with his electric-car business devastated by the success of self-starting gasoline vehicles, Fritchle determined to build a practical, inexpensive wind-power plant for rural use. He wrote to several companies—the Aermotor Company, Flint and Walling Manufacturing Company (Star Windmills), the Elgin Wind Power and Power Company, the Perkins Windmill and Engine Company, and possibly others—inquiring about the horsepower of their windmill models, and if they sold an electric dynamo that could be attached to the windmill.[17]

The response—for Fritchle's purposes—proved encouraging. None of the companies offered an electric dynamo, although the representative of Aermotor Company stated that his company "has for some time been conducting experiments on an extensive scale looking toward the perfecting of an electrical lighting plant to be operated by wind power." The representative suggested that Fritchle "postpone [his] plans for a lighting plant for some time, [as] we

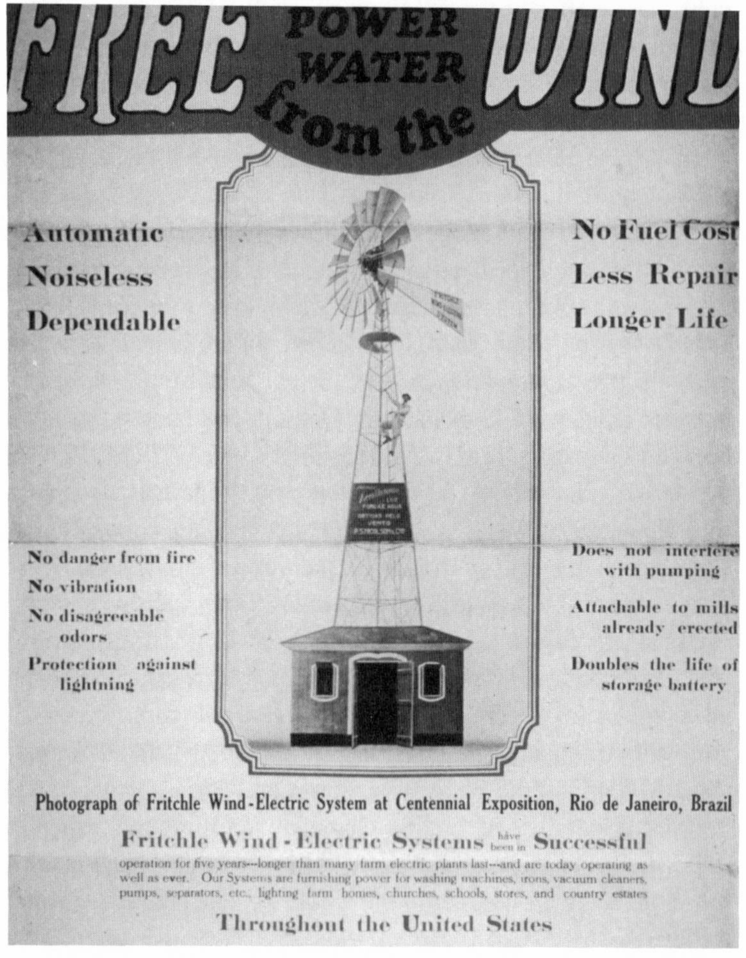

Between 1919 and 1922, the Fritchle Wind-power Electric Company built some fifty-five wind-electric plants. Oliver Fritchle first built electric automobiles but turned to wind generators when self-starting, gas-driven vehicles drove him out of business. His advertisement, "Let the Free Wind Light Your Homes," appealed to farmers, but his simple, inexpensive plant was unsuccessful. (Courtesy the Colorado State Historical Society, Denver, Colo.)

may be able to offer you a very desirable outfit at a reasonable cost."[18] Of course Fritchle had no desire to buy a wind plant; he simply sought information regarding competition from the windmill companies. He was spying. Since the response regarding dynamos was negative—that is, positive from the young inventor's point of view—Fritchle continued his efforts. He did not work on a windmill but devoted his time to patenting a wind-activated generating plant that could be attached to popular water-pumping windmills. With optimism but not much capital, the inventor incorporated the Fritchle Wind-power Electric Company on January 24, 1919.[19]

Company records indicate that Fritchle sold perhaps fifty-five wind-electric plants, primarily in the states of Illinois, Iowa, Kansas, and Nebraska.[20] His sales did not equal his expectation, but Fritchle seemed able to find endorsement letters from satisfied owners. Henry Anderson, of Haxtun, Colorado, noted that with the new Fritchle plant he did not "have to worry about cranking My Silent Alamo Electric Light plant that refused to start the greatest part of the time." No more smelly fumes, only the "free Colorado wind" for the enthusiastic Anderson. "I am in good humor now. I can get along with my family, a little child could play with me."[21] But the endorsements did not bring prosperity. By 1921, Fritchle's budding enterprise was struggling for survival. He came close to striking a deal with the Julesburg Tractor and Implement Company to move and expand his business, but the $30,000 cash he required to produce an anticipated six hundred plants proved to be more than he could raise.[22] Fritchle looked elsewhere, working out an agreement with the aforementioned Woodmanse company, makers of windmills, to relocate his operation to Freeport, Illinois.[23]

Once attached to Woodmanse, Fritchle took what might be considered to be a sneaky step to boost his lagging sales. He rented a post office box in a nearby town, at Kings, Illinois, under the name C. B. Henderson. Fritchle then wrote to his wind-plant customers under the name of Henderson, stating that he was considering purchase of a Fritchle plant. What did they think of the ones they had bought? Most of the responses that have survived were quite positive, although Fritchle may have discarded the less favorable replies. Perhaps Fritchle hoped for a degree of honesty he would not have received under his own letterhead; or perhaps he adopted the ruse because the fictitious Henderson would not be responsible for problems that such letters might reveal.[24] Either way, Fritchle's struggle for survival of his enterprise came to an end in 1922. He went broke—a victim of more advanced technology and new competition.

Much of the competition came from the Perkins Corporation, which entered the wind-electric field in 1922. Based in Mishawaka, Indiana, the company had gained an enviable reputation with its Perkins Solid Wheel windmill, an attractive and reliable unit, and the Perkins Steel, which "once dotted the countryside in parts of the United States and Canada."[25] Now the company introduced the Aerolectric, "a better electric system than has ever been made before by any one." It featured Hyatt bearings and a Westinghouse generator. However, the wind wheel or rotor was identical with that on their water-pumping windmill, with a maximum speed of only 60rpm. Perkins added a step-up gearing system that gave forty revolutions of the generator to every one of the wind-wheel. Although the company maintained that the entire gear-reduction system "runs in a bath of oil," the Aerolectric unquestionably used a dated technology, one better suited to pumping water than generating electricity.[26]

Farmers deciding on an Aerolectric system chose between a 32-volt and a 110-volt system. For cash-poor families, the system inflicted serious strain on their budgets. The smaller, 32-volt package was listed at approximately $1,000; the 110-volt system ran about $1,200 (the variable was the size of the steel tower). Shipping (of a little over two tons) and installation were extra. Sales representatives might offer customers credit, the most liberal being 15 percent down, 10 percent within three months at 7 percent interest, and the remaining 75 percent of total price to be paid within twelve

The caption for this ad read: "Let the Free Wind light your homes"—a common theme in 1920s wind-energy advertising. (Courtesy of the Colorado State Historical Society, Denver, Colo.)

months at 7 percent interest.[27] Although the cost of the Aerolectric was $1,000 to $1,200, company executives hoped that the product's features would convince farm families that it was good value. In this they were assisted by the growing communications network, most notably the mail-order business. In a recent work, historian Katherine Kay Jellison, noted that during the 1920s farm families, especially women, became more aware of modern appliances; "the pages of the Sears catalog were increasingly filled with pictures and descriptions of mechanized washing machines, electric irons, gas cooking ranges, and other modern equipment."[28] A power source being of course necessary, a wind generator might provide it. Perkins advertised that the Aerolectric "will furnish you with an automatic FARM HAND that will save much valuable time and money for you." The Perkins ad gushed: "The AEROLECTRIC is as good as a maid for your wife—it will do the washing—make ironing a pleasure, make cleaning a joy—she can use a toaster, a percolator, curling iron, and any of the many conveniences, which are operated by electricity. No need to use it sparingly, because it is FREE—it's only the wind in another form."[29] A farm magazine, *American Threshermun*, agreed that the Aerolectric made good use of "this mysterious form of energy," noting that the system led the editors to believe that "the modern windmill will be as far ahead of the twelfth century type . . . as the present automobile is ahead of the medieval cart."[30]

THE NEW TECHNOLOGY

But in 1922, the Perkins Aerolectric was not the state of the art. Both Fritchle and the Perkins systems used the traditional, multi-bladed wind wheel—units that were adapted, rather than designed, for wind-electric use. They were never satisfactory. Had a wind plant employed a propeller with high rpms of only two, three, or four blades, the generator could have been turned without an elaborate gear system.[31] Herbert E. Bucklen, an engineer in Elkhart, Indiana, designed the first wind-electric system that took advantage of the propeller blade. He called the unit the HEBCO, using the initials of his name. Bucklen manufactured his units in Elkhart, using a 32-volt system. The propeller was unique and the company literature later proclaimed the propeller to be the same as that that "Col. Lindbergh used . . . in his successful flight to Paris."[32] The propeller system, maintained the literature, proved so superior that it would perform "in either a 10-mile wind or a 100-mile hurricane." Not only did HEBCO take advantage of new advances in aeronautics, the corporation also profited from the new field of aviation

Herbert E. Bucklen was the first in wind-energy history to profit from the aeronautic advances of World War I. Using the initials of his name, he began production of the HEBCO turbine in 1921 or 1922. (Courtesy the Panhandle-Plains Historical Museum, Research Center, Canyon, Texas)

and airmail service. The U.S. Post Office maintained many landing fields. The important fields had high-line electrical service, but many isolated landing strips had batteries charged with a HEBCO wind generator to power their beacon lights. The revolving beacons turned on and off at night, using automatic-sensing units.[33]

HEBCO brochures, like those of every wind-energy company, featured testimonials from satisfied customers. William L. Ayers of Wheatland, Wyoming, praised the unit for its ease of installation.

"I have handled lots of machinery," stated the Wyoming rancher, "but never had anything that ever went together any nicer and perfect in every way as this outfit." Another satisfied owner stated that the unit could generate five times more electricity than his family needed. A father of twelve said he "would not give it up for two dozen of the best gas plants ever built."[34] The HEBCO publicity flyer may have surpassed its competitors in hyperbole. The total yearly expense would be "75 cents worth of lubricating oil" and the electrical current was so controlled that "*no matter what happens to the plant it protects itself*." The plant was so sturdy that "sixty, ninety, one hundred mile winds do not bother the HEBCO." Like other companies, HEBCO capitalized on the idea of free wind. The firm promised prospective owners that they would have "electric power, running water under pressure—all *Without Fuel Cost*—without the noise and bother of a gas engine. . . . The wind is free," proclaimed the flyer, "costs you nothing . . . at the touch of a button you have light or power in your house and barns for milker, separator churn, electric iron, sewing machine, etc."[35]

HEBCO capitalized on aeronautic technology, but it also tapped into farmers' fears of growing dependence on petroleum: they had discarded the horse and embraced the tractor. Everywhere, landowners replaced the intermediate technology of the late nineteenth century with the new mechanization. In the city and on the farm, petroleum-dependent systems replaced renewable fuel systems. Farmers on the West Coast became particularly aware of this dependence in 1920, when a shortage of both petroleum and energy occurred. That summer, a drought created a deficit in hydroelectricity; simultaneously, for four months, petroleum was uncommonly scarce. This brief episode alarmed farmers, who realized that centralization, with all its advantages, was not flawless, and energy self-sufficiency might be more reliable.[36] Other farmers and ranchers felt growing concern at their subservience to the petroleum companies.

LACK OF INTEREST IN ALTERNATIVE ENERGY

According to farmer-philosopher Wendell Berry, they were right to have been concerned. It was during this period, believes Berry, that the "conversion to fossil fuels energy subjected society to a sort of technological determinism." Farmers who once focused on the weather and the biology of living things, now faced technological and economic issues. "Credit, for instance, became as pressing an issue as the weather," Berry wrote in the 1980s, "for farmers had begun to climb the one-way ladder of survival by debt."[37] At the time, the change caused concern in official Washington. In 1924, the secretary of agriculture, Henry Wallace, formulated a question in a memo to the chief of the Bureau of Agricultural Economics, Dr. H. C. Taylor:

> Twenty years ago the power of the farm was furnished mostly by horses and windmills. The horses were fed on the stuff grown on the farm and thus did not require an outlay of cash. Wind was free. All we had to do was to stick a wheel up in the air and harvest it. Now a large part of the power is furnished by the gasoline engine. We have to pay cash for gasoline and oil and for repairs.

Wallace wondered about the trade-offs and if there was a way of "measuring the effect of this change." Taylor's response has not been preserved, but about a year later, T. A. MacDonald, chief of the Bureau of Public Roads, responded to a request by Secretary Wallace for information on wind power by noting that at least five firms were manufacturing "wind electric outfits," and that they had given "fairly satisfactory service over a number of years."[38] MacDonald did not record his source, but he mentioned that agricultural research stations in Iowa, Indiana, and Minnesota carried on research in wind generation. He was not confident of the results, however, for although wind power "affords large possibilities," there was no money for investigations.[39]

MacDonald's lament is a recurrent theme in the story of alternative sources of energy. While government generously supported research and development for central power systems and the "hard energy path" (coal, petroleum, natural gas, nuclear), from the 1920s through the 1960s those companies, inventors, and individuals who advocated alternative energy received no assistance and precious little encouragement.

Across the Atlantic, the climate for alternative energy sources was more hospitable. French companies experimented with larger "propeller mills."[40] More significantly, in England, engineers at Oxford University embarked on an important wind-energy experiment. In 1924, the university's Institute for Agricultural Engineering established a windmill experimental station on the Annables Estate, north of London, midway between the towns of Harpenden and Luton. There, in a year-long, controlled experiment, the engineers tested existing wind machines with the objective of measuring their electrical output. The investigators noted the difficulty of such an experiment: "Whoever sets out to consider the wind and the forces in the air in motion encounters a problem distinct from any other and possessing great complications." The many impediments to inquiry were particularly aggravated when the team attempted "to reduce the vague and apparently aimless vacillations of the air to a really scientific and dependable basis."[41] In spite of the difficulties, the station tested seven machines, representing five manufacturers; and in 1926 reported that the "cost of windmill generated electricity for small lighting and small power purposes was found to be quite reasonable, and as such justifies its wider use in rural districts where there is no general supply."[42]

Noticeably absent from the British tests were American machines. The preeminent industrial nation of the world was totally unrepresented although, certainly, U.S. manufacturers could have profited by participation in the experiment.[43] A Canadian farm journal commented that in electric windmills the British "'out-Yank' the

Yankee."[44] American manufacturers may have been confident that there was no need for them to put their machines up against foreign competitors. American wind machines sold. Companies such as HEBCO probably felt secure in the expanding home market.

In 1929, *Successful Farming* magazine declared: "It appears that the problem of harnessing the wind to produce electricity has been largely solved." The article continued: "There are literally hundreds of farmers in the north central states who are now enjoying the benefits of electric power from the wind." The magazine pronounced that "much of the pioneering period of wind-electric equipment development has been passed."[45]

OLIVER'S BIZARRE "BLUNDERBUSS"

In truth, much work remained to be done in both the United States and Britain before such optimistic reporting would be anything more than wishful thinking. Two unique American-built machines broke with tradition, and with very different results. One machine was bizarre and experimental; the other proved to be a model of efficiency and reliability.

The bizarre machine resulted from the "wishful thinking" of a tall, debonair Texan named Dew Oliver, who promoted real estate in the beach area of southern California. Oliver, according to one Seal Beach resident, could "talk a person out of the gold in his teeth." In spite of his talents, Oliver's real-estate ventures went sour in the early 1920s. Never at a loss for an entrepreneurial gamble, a new idea captured his imagination: on a trip to the Midwest he had observed a number of windcharger machines at work and on his return to California in 1925 he decided to promote the idea. His interest in alternative energy was also boosted when he noticed a fellow in Long Beach generating small amounts of electricity from the ebb and flow of the ocean tide.[46]

Appraising property one day in San Gorgonio Pass, it occurred to Oliver that the wind might be more valuable than the desert land

Dew Oliver built this giant "blunderbuss" turbine in the 1920s on the desert lands of San Gorgonio Pass. It generated electricity but could not forestall Oliver's financial ruin and conviction for fraud. (From *Compressed Air Magazine* 33, no. 4 (April 1928): 2380)

thereabouts. He made inquiries and found that the wind hardly ever stopped blowing, an annoying phenomenon to locals but one that held potential profitability for Oliver. A local Native American tale substantiated what was evident. "When the wind quits in the pass," stated the legend, "the end of the world will have come."[47] Oliver had no data from an anemometer, but his immediate observation, backed by Indian wisdom, suggested that a site ten to fifteen miles east of Banning, near Palm Springs, was the place to build.

Oliver began construction of Oliver's Electric Power Generator, a unique, ten-ton, bell-shaped wind machine that stretched to seventy feet. Sometimes described as like a giant blunderbuss (or what, in today's terms, might be said to look like a huge jet engine), it sat on a circular track atop a concrete foundation. When the wind shifted, the machine turned, funneling the air through a tunnel and compressing it by a factor of twelve. The compressed air turned a series of propellers mounted on a horizontal shaft within the tunnel. Below the tunnel, a generator, powered by propellers, produced two hundred horsepower at peak production.[48] Evidently, the system worked satisfactorily. Much later, Sperry Knighton, Oliver's electrical engineer, said "there was plenty of power from the wind and the idea worked. Only lack of money kept us from powering Palm Springs as Dew planned to do with that turbine. There was no mechanical trouble whatever."[49] However, according to more than one account, the first of the two generators, a relic from an old Seal Beach roller coaster, spun so fast that it burned up. The second installation, a more powerful generator, from the Pacific Electric Streetcar Company of Los Angeles, proved more suitable for the load.[50]

Oliver had announced that his company would build a dozen or more such tunnel-system generators to serve a single electrical district. With that number of units, in dispersed locations, at least one of the units would always be producing. If, perchance, a general calm occurred, a bank of batteries would hold some storage.[51] Simply put, Oliver intended to build the first wind-energy generating station—a decentralized system that would be independent of other forms of energy production.

Oliver did not have the opportunity to prove his claim. Financial trouble—not mechanical—brought his venture to an end. Once the bell-shaped behemoth generated electricity, Oliver formed the Oliver Electric Power Corporation in Reno, Nevada, and capitalized it at $12.5 million. He sold stock in amounts of $50 to $10,000 in the boom market of the 1920s. Investors responded generously.

However, Oliver had not paid much attention to security laws, and in 1929 he faced eleven charges of violation of the California State Corporate Securities Act and grand theft. Tried that April in the Riverside Superior Court, a jury found him guilty on six counts. Judge F. A. Leonard sentenced him to prison for three years, but commuted the prison term to three months on condition "that Oliver abstain from liquor and stay out of places where it was sold."[52] This stipulation probably provides a clue to Oliver's faulty judgment.

It was the end of the road for the wind plant. Once out of prison, Oliver disappeared and there is no record of whether he honored the conditions of his probation. He left a checkered legacy. On the one hand, Oliver must be credited with constructing the first large wind turbine since Charles Brush's effort; on the other hand he offered a dubious example for future wind-energy entrepreneurs. He was a forerunner of the charlatans who, in the early 1980s, would use federal and California laws to put up shoddy wind machines, besmirching the reputation of a struggling industry. Oliver, too, cared more for profits than performance.

To Oliver's credit, he knew where to construct a wind machine. Today, San Gorgonio Pass is one of California's major wind generating stations; Oliver had a good feel for his site. Whether he really had a good design, we shall never know. The long funnel rusted on the desert, the wind moaning the only lament for its demise. It was salvaged as scrap metal during World War II and has never been replicated.[53]

THE JACOBS MODEL

I'm kind of a freak, see. I want things to work forever.
I built my plants to last a lifetime.

—MARCELLUS JACOBS, 1973

Far from California, on the gently rolling prairie of eastern Montana, another imaginative man worked with wind energy. In most respects

Marcellus Jacobs appeared to be the antithesis of Dew Oliver. He was a young, country lad, far removed from the diamond tiepin world of the Texan. His parents had moved the family from his birthplace—Cando, North Dakota—to Indiana and then to the vast, empty grasslands of eastern Montana. Marcellus was one of eight children. Although he finished high school and a year of electrical training in Kansas City and Indiana, he considered himself self-educated. "Most of my education," he later recalled, "just came from studying on my own. I got the books and picked up what I could from them, and thought the rest out for myself."[54]

Jacobs's initial interest in wind energy stemmed from a problem that demanded a solution. Although isolated, the family wanted the convenience of electricity. They installed a Delco 32-volt gasoline generator, but living forty miles southeast of the tiny town of Wolf Point, hauling gasoline took three days and necessitated crossing the Missouri River; so Marcellus and his older brother Joe decided to build a wind generator. Resourcefulness was surely a family trait. Already the boys had built peanut radios, a mechanical fence-post digger, and a propeller-driven machine for traveling over snow. If a new technology existed that would make life on the ranch easier, they were willing to try it. And Marcellus—as he later recalled—"had always been intrigued by the wind."[55]

The brothers first attempted to build a wind plant from a water-pumping windmill, but after about three years of experimenting they realized that the multibladed wheel turned too slowly ever to produce much electricity. In the meantime, Marcellus had learned to fly and he realized that the quick-turning airplane propeller could resolve their problem. Using the standard two-bladed propeller, they encountered problems with vibration, but the problem disappeared when they installed a three-bladed style. The propeller design established, the two young men turned to the problems of speed and pressure. To regulate the propeller's speed during high wind, they developed a fly-ball governor. The fly-ball, automatically

The finest wind generator ever manufactured was designed by the brothers Marcellus and Joe Jacobs. The Jacobs Wind Electric Company produced more than 30,000 units between 1927 and 1957. Here Marcellus admires his turbine. (Courtesy the Jacobs Wind Electric Company, Corcoran, Minn.)

coming into play by centrifugal force, feathered the blades in high winds, both slowing the propeller and relieving the pressure on the blades. It was a system superior to any other.[56]

All of this work took place on the Montana ranch. It was from there that the brothers incorporated the Jacobs Wind Electric Company on a limited forty-year charter.[57] Between 1927 and 1931, Marcellus and Joe assembled about twenty Jacobs wind plants. The plants worked and neighbors noticed—and soon they purchased the wind machines for their own lighting systems. Not only did they buy the plants, they also invested. The ranchers of Montana were the ones first to sink money into this emerging enterprise. It was a worthwhile investment, for the Jacobs boys made a reliable product and ran a conservative business, avoiding loans and insisting on cash payments.

The prairie of Montana proved to be a fine laboratory. The conditions were as severe as any in the country—but it was a poor location for a factory. When demand increased, the Jacobs Wind Electric Company moved. In 1931 the boys left the ranch and Marcellus and Joe moved the operation to Minneapolis, Minnesota. There the Jacobs brothers carried out more technical research. They adopted a larger generator, improved the pitch of the Sitka spruce propellers, and balanced the generator load to match the efficiency of the propeller, essentially allowing for a variable-speed system. After about ten years of research, experimentation, development, and twenty-five patents, they perfected the Jacobs wind generator. By the end of the 1930s, the Jacobs brothers offered a truly reliable unit—although from 1936 to 1956, Marcellus would go on making changes further to increase reliability and efficiency.[58]

Marcellus Jacobs depended on word of mouth to sell his wind plants, although, in the summer, he would also travel to county fairs and small towns. He went by pickup truck, with a wind machine set up in the bed. It was the pickup truck that attracted one of Marcellus Jacobs's most loyal employees—Fred Bruns. Bruns had farmed in eastern South Dakota, but drought and the Great Depression drove him from the land and into the ranks of the unemployed. One Sunday morning in a Minnesota town, the unemployed Bruns spotted Jacobs's truck. He was fascinated with the wind machine and wanted to meet the owner. As Bruns later told it, he determined "to sit here if it takes all day."[59] Jacobs showed up, with the outcome that Bruns installed more than 250 Jacobs wind machines in North and South Dakota, Minnesota, and Canada. He worked full time in the wind-energy business from 1934 to 1942. Mostly, he worked for Jacobs dealers, but he also installed other makes of machines. Aside from HEBCO and Wincharger, however, his experience with other makes (Air Electric and Wind King, for example) was less than satisfactory. The Wincharger "worked after a fashion," but it, like all the others, often blew down in high winds. The Jacobs machine, however, had a good governor, and that fact made all the difference.[60] A number of companies manufactured wind machines, but without question the Cadillac of the available options was the Jacobs.

When Marcellus Jacobs was traveling, he looked for dealers, not customers. Once the dealer had sold a few machines, the primary advertisement was "word of mouth."[61] Satisfied users sold the machines. However, some successes were worth advertising and one company brochure claimed "seven years of perfect operation in all kinds of weather, high winds, storms, ice, snow, or sleet, in many parts of the world, even with Admiral Byrd at the South Pole, is your assurance of permanent, reliable and attention free service year after year." The Jacobs flyer proclaimed: "Wind! The Cheapest Power in the World Is Easily Available To Every Farm Home."[62]

Dependability was the hallmark, and some stories became legendary. A Christian mission in Ethiopia that had installed a Jacobs plant in 1938 finally, in 1968, purchased some replacement parts—their first, a set of generator brushes.[63] Admiral Byrd installed

This Jacobs wind turbine traveled to Antarctica with Admiral Byrd in 1933, providing electricity for the Little America outpost from its seventy-foot tower. (Courtesy the Jacobs Wind Electric Company, Corcoran, Minn.)

a Jacobs wind machine at Little America in 1933. In 1947, Richard Byrd Jr., visited the deserted site and found the Jacobs machine intact, the "blades . . . still turning in the breeze." Eight more years passed—then in 1955 one of 1933 veterans returned to find the blades still spinning. The plant's days were numbered, however, for only ten feet of the original seventy-foot tower was free of ice. The visiting veteran mounted the tower and removed the spruce-wood blades, noting that they showed "little signs of weathering."[64]

When Richard Byrd Jr., son of the explorer, visited the deserted polar site of Little America in 1947 he found the Jacobs still "turning in the breeze." In 1955, one of the 1933 veterans removed the blades because only ten feet of the original tower was free of ice. (Courtesy the Jacobs Wind Electric Company, Corcoran, Minn.)

In an industry in which the technology had a reputation for unreliability, the Jacobs product was powerful and trustworthy. Marcellus Jacobs "built redundancy into the system" no matter what model or what price.[65]

However, the notion that wind energy was available to every farm home that wanted it proved unrealistic. As the "Cadillac of the trade," a Jacobs wind machine was an unattainable dream for many Depression-era farmers. The least expensive model, a 32-volt system (2,500 watts) sold for about $490. The fifty-foot tower was another $175; the 21,000 watt-hour glass-cell, lead-acid storage battery was an additional $365. That made it an investment of $1,030, no small stake during those discouraging years, especially since the company did not offer special credit terms.[66] Beyond the cost of the system, customers had to wire their buildings and purchase 32-volt (or 110-volt) appliances. So financially, a Jacobs wind system was not, in the words of the ad, "easily available to every farm home."[67]

Like most farmer-dependent companies, the Jacobs Wind Electric Company suffered from the lack of a stable market. Even those rural families who could afford the plant might shy away from such a substantial investment if the possibility of central power service loomed in the future. And in rural America, a steady economic decline had preceded the depression of 1929. The stock market crash simply confirmed, rather than announced, the longer, less dramatic decline of U.S. agriculture. Few farmers had the resources for labor-saving technology and as the depression deepened their predicament became worse. For instance, the farmers of the Texas hill country could not even afford the $400 for a water-pumping windmill. In such cash-poor regions, a wind generator was out of the question.[68]

However, reliability kept the company in business. Within a few years the machines had gained an enviable reputation, enhanced by an article in the reputable *Agricultural Engineering* magazine.

Author Fred Hawthorn gave a very positive report on wind energy in general, responding in some depth to the crucial issue of reliability. Were wind machines reliable? His reply: "Based on our three years of practical experience, the answer would have to be an emphatic Yes." Hawthorn's figures relied on a report about thirty Kansas plants that showed an average annual repair and oil bill of $1.56. Another study on sixty-six plants in Iowa found the annual repair bill to be $1.99. Hawthorn used an average of $1.75 and then added depreciation on the plant (twenty-five-year life), the batteries (eight-year life), and the tower (forty-year life), as well as interest on the investment. The annual cost came to $66.15 per year, or approximately $5.50 per month for electricity. His cost per kilowatt-hour turned out to be .055 cents, based on 100kWh per month.[69] These figures were sensational, and although the initial outlay might be high, if this kind of economy could be realized, rural Americans would be tempted.

Hawthorn also emphasized the emerging idea of energy independence, free from restrictions of private power companies or Rural Electrification Administration cooperatives. "The outstanding advantage of wind generated electricity," waxed Hawthorn, "is that its cost is absolutely independent of the customer-per-mile factor. Extreme isolation means nothing to a wind plant owner."[70]

In an effort to keep professional objectivity, Hawthorn did not mention the Jacobs plant. However, he gathered much of his encouraging data from his own plant, and the name *Jacobs* can easily be read on the published photograph of his weather vane.[71] A few smaller, different wind plants are mentioned, but the author favors his own, stating that "personally I recommend the purchase of both large size plant and batteries, as such an installation will carry the load through calms, deliver a far superior, heavy-duty type of service, and will take care of the increased load that will surely come as more gadgets are added."[72] And those who could afford the machine undoubtedly found life easier. Consumers could purchase Jacobs-brand household products—actually manufactured by the Hamilton Beach Company—waffle irons, vacuum cleaners, laundry irons, air conditioners (swamp-cooler type), refrigerators, freezers, and so forth. These were fine products, and it was said that the freezer was so well insulated it could last five days without power.[73] Of course, the wind-electric system also ran electric motors and farm machinery, but the company made its greatest pitch for the transformation of the farmhouse. Because of the profound impact of electricity on the home, Jacobs dealers appealed to the ranch wife, who often acted as the decision maker, pacesetter, and vociferous advocate of electricity.[74]

"USING UP THE WIND"

How many wind units did the Jacobs Wind Electric Company manufacture? We can only speculate: Marcellus Jacobs was quite secretive regarding statistics. He made a number of model changes and after each change he modified the serial-number system. Furthermore, according to his son Paul, the number of units he produced simply was not relevant to his advertising or his sales techniques. Later in life, Marcellus indicated that his company "must have built about 50 million dollars worth of plants in 25 years."[75] If the average cost per plant was $1,500 (an estimate on the high side), his company produced over thiry thousand wind plants.[76] Similarly, we have to speculate about the total number of wind-electric systems sold. Hundreds of thousands operated in the late 1930s and 1940s, tailing off to about fifty thousand windchargers around 1950.[77] They served all over the West, but particularly in remote areas—areas that could not meet the basic population-density requirements of the Rural Electrification Administration. Mainly, they were utilized to contribute to the electrical needs of ranches and farms, but they also served other functions; for example, providing small charges of negative direct current on underground pipelines to prevent deterioration, charging telephone relay stations, providing light for remote airplane runways, running stripper oil wells in Texas, and providing power for solitary lighthouses. Fred Bruns recalled installing a Jacobs machine in Canada at a wilderness mink farm. To get clear of the surrounding forest, he topped four trees at sixty feet, built a platform, and erected a sixty-foot Jacobs tower on the platform. The tree-stump foundation bore the tower well and the machine provided power for many years.[78]

The fifty thousand figure does not take into account smaller wind units. Radio companies (e.g., Zenith) sold small, wind-driven units to charge a six-volt battery. The Sears Roebuck catalog marketed a Silvertone Aircharger in 1946 for $32.50, with batteries from $2.69 and up.[79] The husband-and-wife owners of a store in Fluvanna, Texas, recalled that "we were really 'uptown' when we got 'wind-chargers.' " They only sold a few of the larger models, but hundreds of the $15 and $18 models made their way to neighboring ranches, usually to charge a Band C dry battery and a 6-volt storage battery for the family radio.[80] R. E. Weinig, the general manager of Wincharger Corporation, testified in 1945 that some 400,000 Wincharger plants operated worldwide, of which only about 25,000 were large plants.[81] Wincharger Corporation produced most of these small units. Other brands included the Miller Airlite, Universal Aero-Electric, Paris-Dunn, Airline, Wind King, and Winpower.[82]

The fifty thousand figure also does not take into account those inventive farmers and ranchers who, like the youthful Jacobs brothers, built their own wind-energy plants. Ranchers have always been dexterous and they rarely discard old cars and trucks—a handy source of parts to fashion a wind plant. One west Texas rancher recalled how he combined a propeller, a 6-volt battery, and an automobile headlight to make his wind plant. Another recalled that his family "had good lights" from a little 6-volt system they put together.[83] When the *Nor'-West Farmer*, a Canadian journal, in 1937 offered to provide readers with instructions on how to build a small

(6-volt) wind plant, correspondents swamped the editors with requests. Sending out the instruction sheets, one employee remarked that if everyone who requested the information built a plant, "they would soon use up all the wind."[84]

In recent years, wind-energy devotees have purchased and rebuilt many old Jacobs 110-volt generators. They are considered to be the finest wind plants ever manufactured anywhere in the world. In 1989, *Home Power* magazine continued to recommend restored Jacobs, saying these old turbines have "proven reliable and will remain popular until the demand for new wind generators decreases their price"[85]—a tribute to the work of the two Montana youths who, with little formal training, made it happen.

Although no machine directly competed with the Jacobs, the Wincharger Corporation, of Sioux City, Iowa, delivered thousands of wind units to U.S. farmers and ranchers. The corporation began production in 1935, gaining popularity because their units were of comparatively moderate price. Ranchers called Winchargers the Chevrolet or Model T of wind energy, partly because of its more affordable price, partly because of the range of models available, from 6-volt to 110-volts and from 200W to 3,000W.[86] Often, a rancher hesitating while anticipating central power, and unable to afford a Jacobs, purchased a Wincharger. Most Wincharger models used a four-bladed wheel that operated at a satisfactory number of revolutions per minute. The governor, or brake, however, posed a problem. Unlike the Jacobs, the Wincharger had fixed blades, so although the brake slowed down the propeller, the pressure remained. In a gale, the pressure often became so great that the apparatus was wrecked by being blown against the tower. Bruns, who installed a number of Winchargers, said "they worked after a fashion," but "often blew down."[87] Later models, somewhat remedied this flaw, utillizing a variable-pitch governor that in high winds could feather two of the four blades.[88]

The Wincharger did not gain the reputation of the Jacobs, but it had many satisfied customers. For example, Ida Chambers and her son, Roy, supplied their Jackson Hole, Wyoming, ranch with electricity between 1946 and 1954 with a Wincharger 32-volt system. In an interview, Ida Chambers fondly recalled that the machine required no repairs and that it was a "very, very good investment . . . the greatest thing that ever was."[89] Her daughter-in-law, Becky Chambers, remembered differently. When Ida hooked up to REA power, Becky recalled, she "gathered all her 32-volt appliances, marched out to the yard, dug a hole, and threw them in—glad to be rid of them."[90] However, the Chambers would have continued to use their machine but for the desire for television and the like (unavailable on a 32-volt system). Although the Chambers family moved and the ranch is now part of Grand Teton National Park, the Wincharger remains, a little the worse for wear but still atop its tower—and Jackson Hole, although not subject to the severe winds of the Great Plains, is not known for mild weather.

The Jacobs Wind Electric Company went out of business in 1956. That same year, Wincharger ceased production of all but a tiny, 200-watt miniature machine. Marcellus Jacobs said he just quit trying to fight the tough combination of AC power and the REA.[91] In truth, when Jacobs first perfected his wind machines, and when the first Wincharger came off the assembly line, even greater forces were at work to limit the potential of wind energy and eventually destroy the market: centralized systems, with all their ramifications. Marcellus Jacobs recognized the trend and in an attempt to adapt he proposed to the U.S. Congress in the mid-1950s that his company provide "high line grid system booster A.C. wind electric plants." The Jacobs Wind Electric Company would build and install, on grid towers, one thousand wind generators, one mile apart, on the power-line system from Great Falls, Montana, to Minneapolis, Minnesota. The wind-generated energy would feed

Modestly priced Winchargers were the most popular early units. This one powered the Chambers homestead on Mormon Row, Jackson Hole, Wyoming—the Teton Mountains forming the backdrop. Many farmers could not afford Jacobs models—the Cadillac of the wind machines. (Courtesy Paul Gipe and Assoc.)

into the grid. Neither Congress nor power-transmission executives responded favorably to the idea.[92]

In the grandiose plans of the power companies, there was no niche for small, wind-energy systems. Private power companies proved to be the first nemesis of a neophyte industry struggling to perfect its technology and raise capital. Private power, however, left plenty of possibilities for wind-energy use on ranches and remote farms. The Rural Electrification Administration closed this remaining window of opportunity, as we shall see in the next chapter.

CHAPTER 5

THE REA OFFERS A DIFFERENT WAY

It is natural that the owners would not want to junk their [independent] plants and join the high-line immediately. This is particularly true of the wind-driven plants which have low operating costs.

—RURAL ELECTRIFICATION SURVEY, KANSAS STATE COLLEGE, 1940

NO ONE WOULD DENY THE SUCCESS OF THE RURAL ELECTRIFICATION Administration (REA) in performing its task. But neither would anyone dispute that the REA sounded the death knell for small wind-energy systems. As already noted, Marcellus Jacobs, for one, simply became weary of fighting the REA. However, the REA was only the final blow in the battle over providing power in both the city and country. This struggle eliminated the market for individual systems of power production and nipped in the bud the concept of cogeneration, whereby companies that produce heat or steam generate electricity for commercial use. The wind-electric companies were hit hard by centralization and by government subsidies for central systems. No wind-electric company survived.

Thomas Edison took the first step in centralization in 1882 when he built New York's Pearl Street generating station. From that time on, U.S. capital and inventiveness strove to create huge, centralized systems, in communications, highways, and power. According to one historian of technology, the genius of U.S. invention in the late-nineteenth century and the twentieth century was not in the introduction of specific inventions such as the telephone, electric lighting, and the automobile, but rather in the development of systems to utilize such inventions.[1] In retrospect, if Charles Brush had designed a synchronous converter to feed his wind-dynamo energy into a central grid system, he would have made a more significant contribution to American invention.

Where Brush failed, Samuel Insull succeeded. If the wind-energy companies had a nemesis, recognized or not, it was this Chicago entrepreneur. Insull's genius was at amalgamating electrical systems. In time, his Middle West Utility Company owned or controlled a vast utility empire from one end of the nation to the other, representing assets of over $1.2 billion. At the height of his power, in the 1920s, Insull opposed any decentralized systems or individual plants. Constantly, he declared his faith in private ownership of power production and distribution. He championed what historian Harold Platt has called "the gospel of consumption."[2] In Insull's scheme of thing, gasoline and wind-electric plants in rural areas were no more than tiny gnats, but they deserved to be swatted. For utility companies committed to complex electrical grids, any independent, small-scale power plant represented an annoying anomaly: Disorder within an orderly system—a problem, not a solution.[3]

Samuel Insull's initial interest in rural electrification can be dated to 1906, the year he purchased Hawthorne Farm, a country estate north of Chicago. On weekends, Insull motored to his rural retreat to enjoy the pastoral life. Although he appreciated nature, he also cherished the comforts that the city—and electricity—afforded. Within a couple of years, Insull had upgraded his personal electrical service, and, in the process, that of the county. He enthusiastically embarked on what became known as the Lake County experiment—an effort to "electrify" the countryside.

This first rural electrification project proved successful, both in terms of profit for the utility company and service to the rural consumers. Other utility companies followed suit, running lines to villages and rural residents. By 1920, according to Insull's biographer,

> the systematization movement had gained such momentum that—along with the new automobile industry—it would dominate, and then carry, the American economy for better than a decade. By the time it was over, electric power would be so abundant and cheap in the United States that people who had never expected to use it found it as natural and as necessary as breathing.[4]

Such an appraisal of Insull's role in rural electrification might lead one to believe that Insull, the energy czar, and his fellow executives in utility companies busily initiated programs, running countless lines to the hinterlands of America. This was not the case. Insull was sympathetic to rural electrification, but he had not risen to power through good works and philanthropy. Shrewd business practice called for caution and Insull would not construct a rural electric system unless he believed it would be profitable.

However, shrewd business practice also called for sympathetic interest. The National Electric Light Association's Committee on Electricity in Rural Districts (primarily a Insull-inspired construct) provided the vehicle for both sympathy and caution. After studying the situation, the committee's 1913 report noted the need for rural electrification, but also suggested that apathy among farmers hindered progress. The report implied that delays and inaction resulted from the paradoxical attitude of the rural community. While many embraced progress, others were indifferent; farmers lacked agreement. In a situation of client indifference, inaction and delay seemed to be the most prudent course. The low level of interest extended to federal government officials, too. When Insull's committee approached the secretary of agriculture, James Wilson, seeking federal assistance for rural electrification, "they were greeted with an enthusiastic yawn."[5]

THE GRID GETS GOING

Following World War I, the utility industry enjoyed a decade of tremendous expansion and achievement, "the greatest the power companies have ever known."[6] Cities and towns across the country installed central electrical systems and industry provided a stream of new appliances to consume the product. As part of this growth, the utility companies, the federal government, and farm organizations shook off their lethargy, recognizing the social and economic importance of electricity on farms and ranches.

However, the National Electric Light Association (NELA) made little progress in satisfying the demand for energy. In 1924, the Committee on Electricity in Rural Districts reported that "the development of rural electric service from central stations has just started."[7] The committee professed an interest in rural electrification, and Insull and other leaders spoke and wrote of the need and their efforts to fill that need. But it was a smoke screen. Why lose money in the country when the utility companies could prosper in the city? And so the utility companies created an illusion of interest in solving the farm problem.

Statistical evidence reveals that there was only sluggish progress. In twenty-seven states, only 300,000 farms and ranches were hooked up to central power.[8] Nationally, statistics reveal that of approximately 6,371,000 farms throughout the nation, only 500,000 were part of centralized electric power.[9] In the 1930s, Rural Electrification Administration bureaucrats usually gave a 10 percent figure when quantifying the sad state of electrical service in rural America.[10]

Accounting for such an abysmal record, Insull and others blamed the cost of stringing power lines. The cost in rural regions ranged from $800 to $2,400 a mile, depending on labor rates, terrain, soil, and other factors—figures the companies often found to be prohibitive. The Wisconsin Power and Light Company, figured $1,405.12 per mile as their average. Using such a figure, service for a farm valued at, say, $8,000 and located six miles from a line would cost more than the value of the property.[11] Insull had no ready solution for this problem. He certainly did not want government involvement, but neither did he suggest that individual electric-power plants might be the answer in isolated regions. He talked of cooperation and community, using an urban example: "Our manufacturing industries have been enabled either to junk their old power plants or to construct factories without installing equipment for generating power." Rural America could do the same. "This will happen in similar fashion on the farm [from] the pooling of the power supply of a great number of consumers [thus] the farmer's capital costs for power are reduced." The answer, Insull believed, lay in cooperation. Appealing to traditional American values, he called for town meetings in which a group of rural neighbors might "exact every advantage over the utility companies." Insull talked about "man-to-man" conferences and "intelligent compromise"; it would all be for the greatest good for the greatest number.[12] Of course, the power company held all the aces in such bargaining. Although the company might bend a bit, the bottom line remained: profit for the stockholders.

The private power interests formed impressive political and economic groups; for example, the Committee on the Relation of Electricity to Agriculture. This committee represented NELA, the American Farm Bureau Association, the American Home Economics Association, the American Society of Agricultural Engineers, the National Association of Farm Equipment Manufacturers, and a number of other trade organizations. Also included on the committee were the Departments of Agriculture, Commerce, and Interior.[13] Such a powerful combination of economic and political interests should have resolved the issue of rural electrification, but it could not do so. Although Insull and others proclaimed in 1930 that "rural electrification has been made possible by the rise of the great utility investment companies," the statistics tell a different story, revealing a utility monopoly that was indifferent to public interest.

This monopoly did not go unnoticed. During the late 1920s, even as Marcellus Jacobs, was working toward perfecting his wind-electric plant, the utility power brokers came under increased attack. Well known leaders such as Gifford Pinchot, then governor of Pennsylvania, proclaimed that "there is an electric power monopoly." That monopoly, he said, is organized "for ruthless exploitation, uninterrupted and unrestrained by anything approaching effective Government intervention and control."[14] Recognizing the growing importance of electricity, Pinchot predicted that "before very many years it will be not 'the hand that rocks the cradle,' but the hand that turns the electric switch that will rule the land." Pinchot pointed out that Insull controlled not only the Middle West Utilities Company but five other companies, too—a total service area of more than twelve million people.

To expose the domination of the power business by the utility holding companies, Pinchot organized his Giant Power Survey, an investigation into the policies of holding companies in Pennsylvania. To lead the investigation, Pinchot picked Morris Cooke, an energetic engineer who today would rate the description of social activist. Cooke took the job seriously, and although he failed to entice private utility companies to serve rural areas—and to charge reasonable rates—he did gain experience that served him well when later he became administrator of the Tennessee Valley Authority (TVA), and then head of the Rural Electrification Administration.

In this vast power struggle over electricity supply, the independent wind plant became one of the victims. New Deal TVA chief David E. Lilienthal noted that early in the century "the supply of electricity began as a local enterprise. Throughout the country, separate electric generating plants and distribution facilities were set up in each community."[15] Lilienthal was referring primarily to towns, but the same could be said of rural America. The first rural electric service consisted of independent plants—well over 300,000 of them by 1930.[16] But, like Lilienthal's small-community power systems, they gave way to the great engineering developments and what Lilienthal called "the financial wizards." Efficiency emerged victorious, but in certain respects the people lost.

Independent electrical systems, although clearly inferior to central systems in capacity, served thousands of rural families. Although many might have agreed with Clara Jensen, a Wyoming ranch wife who was mighty pleased to be rid of the old windcharger because it was always breaking down, an equal number found that central power did not create an electrical paradise in the countryside.[17] Central power did not preserve the family farm; nor did it halt the drift to urban centers. Historian Katherine Kay Jellison maintains that, in fact, electricity and new technology drove farm families from the land, since "fewer . . . were actually needed on the land." Those that did remain found that to pay for the increased investment in technology, the women often "used the family automobile to travel to wage-paying town jobs."[18]

1935: ENTER THE REA

Into this situation of vacuum—with only scattered use of individual power systems and failure on the part of the private utilities to provide rural electricity—stepped the federal government. In 1935, President Franklin D. Roosevelt's New Deal administration created the REA, acting under authority of the Emergency Relief Appropriation Act. The REA's mission was to "initiate, formulate, administer and supervise a program of approved projects with respect to the generation, transmission and distribution of electric energy in rural areas." In that same year, public fear of the utility holding companies resulted in passage of the Public Utility Holding Company

In rural areas, companies aimed their advertising at both the farmer and his wife. The Wind-Power Light Plant, of Des Moines, Iowa, specified the many tasks its wind turbine would perform on the farm. (Courtesy T. Lindsay

Act. This measure reined in the companies, limiting profits and lowering rates. The two pieces of legislation brought some relief and a ray of hope to rural Americans.[19]

Roosevelt named Morris Cooke—a person superbly qulified for the task—to head the REA.[20] Cooke quickly convened a formal conference of sixteen leading utility executives and asked for their cooperation in providing electricity to farmers. After a three-month study, these executives reported that "there are very few farms requiring electricity . . . that are not now served." Their recommendation was that the government "consider the immediate urge for rural electrification as a social rather than an economic problem."[21] Where an economic problem might exist, the report suggested, the federal government should subsidize rural power lines. These lines, however, would be owned by the local utility company, which would determine the rates.

This was not what Cooke had hoped to hear. Politely rejecting the offer, he ended the brief honeymoon between REA and private power. Intermixed with the technical discussion there lingered a political perception that as long as That Man Roosevelt continued in office, the utility CEOs would not cooperate.[22]

In the months that followed, Cooke and his engineers and advisors made progress. First, under the leadership of Senator George Norris they convinced Congress to pass the Rural Electrification Act, which Roosevelt gladly signed into law on May 20, 1936. The act not only gave the REA permanence, but opponents could no longer argue that REA was simply the creation of a presidential decree, lacking the support of the legislative branch. Second, Cooke and his staff hammered out a modus operandi for the agency. Farmers would be encouraged to organize cooperatives and then apply for low-interest federal loans to buy equipment and string the lines. These farmer-cooperatives would have to comply with certain standards. One of the most important involved density. A Canadian interested in rural electrification, Sir Adam Beck, premier of the Province of Ottawa, had determined that rural electrification was feasible with a density of three farms per mile of line. Cooke and his assistants adopted this standard.[23] Obviously, in the sparsely inhabited West—and not only the West—this rule would leave much room for independent wind-electric plants.

Meaantime, the manufacturers of wind and gasoline electric plants were standing by, accepting their fate and hoping for scraps of business in isolated areas. But the private power companies were not so compliant: they disputed and occasionally destroyed the new public effort. Organizers, paid by the companies, ridiculed the farmer-cooperative efforts, and occasionally a power company would hastily construct a parallel "spite line" to compete with (and ruin) an REA line. But the REA survived and prospered, as farmer-cooperatives, loans, and expertise brought electricity to much of rural America.[24] It was a social revolution.

But there were also misrepresentations and injustices. REA officials invariably neglected to note the presence of independent electrical plants on the nation's farms. Morris Cooke hardly acknowledged their existence. When quizzed by a congressman regarding his definition of central service, the REA head noted that "Delco is not included"—meaning, by Delco, the company that manufactured wind and gasoline independent power plants. Cooke believed Delco would not object to the stringing of high lines, because the company favored rural development. "If service is established," maintained Cooke, "then the Delco people can come in and make almost as much in selling refrigerators and other apparatus that uses high line service as selling their own plants."[25] Cooke's explanation implied that Delco should do some restructuring, essentially quitting the power-producing business to concentrate on power-consuming products. Companies that manufactured only power-producing units would of course be out of luck.

Neither Cooke nor any other REA official addressed the reality of displaced technology. They would argue ad infinitum as to whether service should be provided by private companies or public agencies, but it was always within a context of dedication to central service. That was not questioned. REA engineers and cooperative officials constantly wrote about the "electrified" farm, confining their definition to central power. Private and public power administrators employed the term *electric service*, implying that bona fide electricity must be transported over lines from a central power source. They seemed to forget, or perhaps deny, the reality implicit as early as 1926 in an article in *Agricultural Engineering*: "Notwithstanding the tendency of most people to define rural electric service as a means of extending . . . electric lines, it must be borne in mind that the individual electric plant has been the means of introducing electric service into many thousands of farms in the United States by providing energy for lighting and small power purposes."[26]

Another criticism of the REA—which seemed to be above reproach—focused on its claims to democratic process and localism. REA officials undoubtedly embraced the idea of local control, but the reality was often different. For example, the loan contract, a document that shaped the local cooperative's rates and policies, underscored farmers' growing dependency on the federal government. And the required standard of three hookups per mile of line, set in Washington, made it evident—according to a recent study—that although it was "designed to emphasize local organization of democratic cooperatives," farmers were nevertheless "brought into closer contact with Washington bureaucracy than ever before."[27]

Without question, thousands of farm homes welcomed REA power as an unmitigated blessing; it was a form of freedom from a world of drudgery.[28] But, viewed from another perspective, central power merely added another nail in the coffin of debt and dependency that has plagued the American farmer throughout the twentieth century. As historian David E. Nye points out, "neither big government nor advanced technology was necessarily consistent with maximizing democracy or preserving the small farmer, and rural electrification certainly would not recognize the agrarian dream of a decentralized blend of technology and Jeffersonianism."[29]

For all the convenience and efficiency it brought, rural central power assisted in a dramatic displacement of people. In 1910, half of the nation's population lived on farms, by 1950, this figure was one in seven; by 1985, only one in thirty-three remained on farms.[30] Wind-plant companies were fond of advertising that the plant would be "the hired hand" and this proved to be even more true of the energy offered by the central system. The REA might be said to have marked the passage of the small farmer and the emergence of agribusiness. The national commitment to a hard energy path (central power produced by fossil fuels) appeared to lead to an energy cornucopia, but that pathway included pitfalls that were not evident in the late 1930s to the supremely self-confident, dedicated officials of the REA.

For those committed to central power, the path seemed to be so straight and true that they sometimes put pressure on unpersuaded farmers. Many regions did not qualify for the three-hookups-per-mile rule, but even when they did some farmers were reluctant to get aboard. Often these were farm or ranch owners who already had an independent generating system. Robert Brown, a farmer-rancher from Hershey, Nebraska, recalled that with his wind plant "we used to generate all the electricity we needed." In 1946, the REA cooperative extended its lines to his place and wanted to hook him up. "We said no," said Brown, "but two years later they came back and we finally said yes." Certainly, a factor in his decision was the attraction of trouble-free electricity. But so was the pressure of the cooperative.[31] Fred Bruns, the farmer-mechanic who installed more than two hundred Jacobs plants, was another to recall the use of pressure on wind-plant owners.[32] Paul Jacobs, son of inventor Marcellus, believes that intimidation continues to this day. In an interview, Jacobs claimed that utility companies fear "wind energy would run them out of business."[33]

In fact, REA investigators had to determine if there were enough homes to justify a loan and it does not take much imagination to understand that some local REA advocates manipulated and influenced families, pressuring them to sign up. This must have often required a degree of salesmanship and persuasion, for if a rancher hooked up to REA, he would have to disconnect his wind charger.[34] Clyde Ellis, for many years the head of the National Rural Electric Cooperative Association, a political lobby group, admitted that obtaining a right-of-way from property owners was sometimes difficult, "but usually the combination of neighborhood pressure and tact got the job done."[35]

In the 1990s, it is difficult to find evidence that independent energy plants existed and that the owners of these plants did not always welcome REA central power. However, in 1940, a Kansas State College rural electrification survey of two counties (Harvey and Dickinson) revealed that many farmers who relied on their own electric plants were not interested in forming REA cooperatives. Approximately 175 private units were operated in Harvey County. This number included 3 110-volt plants, 65 32-volt plants, 106 6-volt plants, and 1 12-volt plant. Dickinson County tallied a total of 215 plants "of various kinds."[36] The survey gathered a lot of material. Although most of the statistical data can not be used to make direct comparisons between central power costs and independent units, the study did figure the cost of operation of a 1,000-watt gas engine-driven plant and a 1,000-watt wind-driven plant. The engine could out-produce the wind plant, but the wind-energy plant cost only $7.39 per month; the gasoline engine cost more than double, $16.34. Significantly, the survey found that "there seems to be no great desire for electric service on the part of many farmers in Harvey County." Part of the reluctance stemmed from the hard times of the depression, but other farmers were satisfied with their independent systems. The report noted:

> The ownership of a private plant indicates a favorable attitude toward the use of electricity and those who have such equipment are considered good prospects for high-line service. But these plants have in many cases proved to be an obstacle to community action in securing high-lines. Owners of private plants, often the better farmers, have money invested in their plants and are in general well pleased with the service they are getting. It is natural that the owners would not want to junk their plants and join the high-line immediately. This is particularly true of the wind-driven plants which have low operating costs (if the initial cost, depreciation, and interest are ignored).
>
> A few instances have been found where, under the impression that the service would be cheaper, the farmer preferred to purchase his own plant rather than join the high-line.[37]

The evidence of this report is that rural people were far from being of one accord regarding central power. Many people found that their wind-electric systems provided both functional value and energy independence, two traits with deep roots in rural America. Perhaps some of the farmers in Dickinson and Harvey Counties derived psychological benefit in turning a negative environmental force—the wind—into a positive economic factor. Creating energy from the wind can effect a ranching family's attitude toward what has been known to be a disturbing phenomenon.[38]

THE PARADOX

Certainly, throughout the first half of the century many people, in the United States and elsewhere, exhibited what one historian called "a distaste for rapid progress, without rejecting industrialization."[39] Historian and antimodernist Henry Adams had somber misgivings about the accelerated rate of energy use; Scotland's Lord Kelvin, the first scientist of renown to advocate wind-energy use, feared the dissipation of energy supplies would lead to the downfall of civilization. And farm parents may have viewed central power as an extension of the corrupt city: electricity was just one more lure to snatch away their children.

Like other groups, the Dickinson County and Harvey County families held opposing and paradoxical views. Some were satisfied with independent energy plants; others welcomed REA power as a means of joining consumer society; and some fell in between. Historian Don Kirschner caught the ambiguity of such beliefs. In a book about the 1920s, he writes:

> Farmers had been walking a narrow line between their values, which led them in one direction, and their materialistic appetites, which pulled them in another. Their values committed them to the natural, the simple, and the serene; they exalted the independence of rural life and the primacy of things spiritual; they condemned the city as the source of all evil.

And yet—and here is the paradox—"they had been enthusiastic about the technological innovations." Kirschner believed that U.S. farmers of the post World War I era "no longer walked the line between their values and their material commitment; they had crossed it, they had made their choice."[40] Reasonably prosperous farmers such as many of those in Dickinson and Harvey Counties had already crossed the line to the intermediate technology of a wind-energy. For them, that—for then—was far enough. The Kansas State College survey concluded that "complete electrification of the farms in Harvey or Dickinson County does not appear to be practicable under present circumstances."[41] The report did not suggest that the independent farmers should be encouraged, subsidized, or assisted in any way: the "uneven playing field" would become evident to recalcitrant farmers in time and REA would have its way.

Given the support of farmers such as those mentioned in the Kansas survey, it is surprising that the manufacturers of small power plants did nothing to oppose the invasion of the REA into their market. As noted earlier, the utility companies bitterly fought the cooperative movement; wind-energy companies did not. For Marcellus Jacobs and Samuel Insull to have united in fighting the REA would have requried the making of impossible compromises. Insull stood for monopoly, Jacobs for individualism. One championed systems consolidation; the other favored self-sufficiency and decentralization. Even if they had found common ground, the private utilities would have little entertained ideas from a tiny industry struggling for survival.

In truth, the wind-energy industry had much in common with the REA. Both had the welfare of rural America at heart.[42] Both wished to restore, as David Lilienthal would say about REA, "democratic control over electricity."[43] They disagreed on the means to that end. Wind-plant advocates favored individualism and a decentralized blend of technology, whereas the REA pushed for democratic cooperatives and centralism.

LOSING OUT IN CONGRESS

I believe . . . it would be more economically feasible to use individual electric generating plants rather than running powerline systems to these remote farms.

—R. F. WEINIG, WIND PLANT MANUFCTURER, BEFORE A CONGRESSIONAL COMMITTEE, 1945

Perhaps more serious to the survival of the wind-energy business was a lack of group organization or cooperation. Like the equipment they sold to farmers, each company stood alone. They had no trade association to deal with common problems. They had no lobbying apparatus by which to argue their case. Thus, the wind industry did nothing to question the objectives or methods of the REA cooperatives. They ceded the main meal, hoping for a few scraps.

Service to isolated ranchers and farmers represented one way wind energy and the REA might have worked in harmony. When, in 1945, R. F. Weinig, vice president and general manager of Wincharger Corporation, appeared before a subcommittee of the House Committee on Interstate and Foreign Commerce he had in mind salvaging a few scraps. He represented Wincharger Corporation, but since no wind-energy trade association existed he claimed to be also the self-appointed, unofficial representative of his industry. Weinig conceded to congressmen that high-line power gave superior service; however, he contended, in some areas it simply was not feasible. Using REA data, from engineers Robert Beale and D. W. Teare, he said 2.5 or 3.0 farm families per mile were necessary to make a cooperative REA venture successful. Weinig quoted statistics that showed that in seventeen states west of the Mississippi River, more than one million farms and ranches were in areas where such operations averaged only 1.1 to the square mile. "There are 500,000 farms that average only 0.6 farm to the square mile," he noted, "less than one-fourth of the minimum required density."[44]

Weinig suggested that families on such farms had three choices: they could remain "unelectrified"; they could be given REA lines, but the district would not be self-supporting and would place a burden on more populous areas; or they could be encouraged to install individual plants. Weinig, of course, favored the latter. "I believe," he said, "that it would be more economically feasible to use individual electric generating plants rather than running powerline systems to these remote farms."[45]

Weinig provided data showing that the more powerful Wincharger then in production would provide one hundred dependable kilowatt-hours per month—ample for the average farm. He did not envision an all-electric farm, but said the wind generator would provide electricity for efficient lighting and stationary rotary engines. Natural or bottled gas would be necessary for cooking, heating, and hot water; a tractor would operate any heavy motors. Weinig presented statistics showing wind plants to be both dependable and economical and noted that the Canadian government had provided loans to rural residents for wind plants. Weinig suggested that H. R. 1742 (the rural electrification legislation) be amended so that the REA program would provide ten-year, low-interest loans to farmers and ranchers for individual energy plants in sparsely settled areas.[46] Under questioning, Weinig noted that most wind plants operated in the western United States; he also maintained that the maximum loan period allowable under the Federal Housing Authority was three years.

Perhaps the most interesting question at the hearing came from Congressman Hinshaw, who wanted to know how, if the proposed amendment became law, the REA would handle the situation? "Would you propose," asked Hinshaw, "that the REA own and rent the equipment or that they merely finance the ownership of the equipment on a chattel mortgage?" Wisely, Weinig avoided the question. It would require considerable study by the government, he said. Both he and Hinshaw agreed, however, that, whatever system the government supported, a maintenance service would be needed to assure that if a wind plant broke down it would be promptly repaired.[47]

Weinig made a respectable presentation and gave competent answers to the congressmen's questions, but his efforst were in vain. Legislators abandoned the amendment. Essentially, Weinig could find no support among the nation's energy brokers. The Rural Electrification Administration, the Tennessee Valley Authority, and the utility companies were wedded to central power systems. They had no patience (not even a frame of reference) for anything so different and—to them—primitive.

In neither manufacturing literature nor magazine articles have wind plants been represented as competitive with central power.

Even the most laudatory article on wind-energy plants, published in 1938, began with the following admonition: "High-line service is recommended wherever it can economically be made available. Wind electric plants are in no sense to be considered competitive, for their field logically commences where the high line ends."[48] Weinig, of course, assumed that there would be an end to the high line—somewhere remote, out there in America. He believed that the REA would abide by its three-to-a-mile rule. It did not: politics intervened. For rural congressmen (Lyndon Johnson is one example) bringing central REA power to their districts represented a victory that would ensure reelection. Johnson would have moved mountains to get REA, and trampling the three-to-a-mile rule proved no great task. He was not alone. But if the high line did not end, no service area remained for wind-energy plants.

Other significant factors worked against Weinig's proposal. At that point in the nation's energy history, leaders pursued the potential of nuclear power and even more complex systems of distribution. The tiny Wincharger and Jacobs plants had no place in the grandiose plans of the REA. In such a climate, giving loan-subsidies for individual energy plants could not be seriously considered.

Thus, for a series of reasons, the wind-energy business slipped again into the doldrums. Whereas, in 1946, the Sears Roebuck catalogue offered the Farm-Master wind plant in either 500W ($98.80) or 1000W ($176.00) models, as well as towers, batteries, and a little Silvertone Aircharger for a radio, in 1947 the company discontinued all such products.[49] In the 1950s, the major wind-plant manufacturers went out of business. The Wincharger Corporation ceased production of all but a tiny plant in 1956; that same year the Jacobs Wind Electric Company closed its Minneapolis factory. The industry lay dormant, a victim of the REA and its own inability to fight for a share of the market.

By the close of 1948, 4,388,200 farms and ranches had received central electric service: 51.2 percent of these were served by power

Many Jacobs wind generators are still in operation today. West Texan Joe Spinhirne (just visible to the left of the unit) stands beside his Jacobs in the mid-1980s. It provided all the electricity for his farm. (Author photo)

companies, 43.7 percent by REA cooperatives, and 5 percent by municipal or other government agencies. This figure represented 75 percent of the nation's farms and ranches.[50] Many farms and ranches remained beyond the high line, but their owners anticipated eventually receiving service. A 1947 article in *Successful Farming* lauded the success of wind-energy plants in North Dakota, but the article also implied that wind power was second best: a stopgap measure.[51] By 1950 a *Successful Farming* article was telling readers not to tear down the wind-plant tower, for it could be put to good use for a television antenna, the communications marvel that was not available with a 32-volt electrical system.[52]

A few farmers and ranchers held on. When, in 1940, REA cooperative officials had approached Joe Spinhirne, a West Texas farmer with a spread west of Amarillo, he felt the high-line offer was too expensive. When his neighbors hooked up, he bought their obsolete 32-volt appliances at dirt cheap prices; he even purchased a discarded Jacobs plant for $100. In the mid-1980s, Spinhirne was still generating his own electricity with the reliable Jacobs. He had become a sort of cultural icon, a folk hero, because of his determination to remain independent. In 1982, one writer labeled him a crafty independent, gifted with "uncanny foresight."[53] In truth, his independence reflected neither philosophical nor ecological principles, but economic rebellion. He continued to refuse to pay the local REA cooperative's hook-up fee, which he considered to be oppressively high.[54]

Spinhirne is an exception. Although he represents the spirit of those Kansas farmers in Dickinson County and Harvey County, such hold-outs are rare. Most rural Americans found the attraction of high-line, reliable power irresistible. Independent wind systems could not match the temptations of modernism. The demise of the Jacobs company and the severe scaling back of Wincharger signified the drift of the times.

CHAPTER 6

IN THE DOLDRUMS, BUT PUTNAM AND THOMAS EMERGE

The great wind-turbine on a Vermont mountain proved that men could build a practical machine which would synchronously generate electricity in large quantities by means of windpower.

—VANNEVAR BUSH

ENGINEERS PAID LITTLE ATTENTION TO THE DEMISE OF THE SMALL, individual wind-electric systems; the turbines, after all, were so diminutive that their loss was inconsequential. Engineers are fond of thinking in terms of scale: If Marcellus Jacobs could build a wind plant to serve a ranch, could an engineer build a larger unit to serve a town? Could a series of these large machines serve a city? And if it is feasible to construct a series of wind generators, could their electricity feed into existing power networks? In short, could American engineers build a large and sophisticated wind turbine? In the 1930s, a young engineer named Palmer Cosslett Putnam asked these questions and did his best to answer them by building a huge wind plant, 1.5-megawatt (1500kW), on the top of a Vermont mountain known as Grandpa's Knob.

However, Putnam was not the first to build one of what engineers presently term Large Wind Generator Systems (LWGS). In 1933, near the windswept Balaclava region of the Crimea, Russian engineer

V. N. Krasnovsky directed the construction of a 100-kilowatt windmill power station. The wind plant featured a three-blade wheel, thirty meters in diameter, that Krasnovsky placed on a tower twenty-five meters off the ground. The engine room, or nacelle, measured thirteen meters in length and could accommodate two operators. The nacelle and rotor sat on a large, rotating, ball-bearing wheel. The Russian engineers began generation of power at twenty revolutions of the propeller per minute. After a few successful months, they increased the rpm to thirty.[1]

Aside from its dimensions, the Balaclava wind turbine's uniqueness lay in a synchronous converter, which allowed the direct current (DC) to be converted to alternating current (AC). Once that was accomplished, the three-phase current, at 220 volts, could be fed into the central system. Thus, the wind plant had no standby gasoline; engine nor did it have a battery-storage system. Soviet Union engineers had solved one of the most difficult obstacles to the practical use of large wind plants—how to integrate them into central grid systems.[2]

By 1935, the Russians had operated the wind plant successfully for two years. The *Electrical Times*, of London, reported that "the Russian engineers are so pleased with their actual working experience of this windmill, extending over two years, in parallel with the local Crimean network, that they are designing plants of greater capacity on similar lines."[3] Evidently they planned to erect multiple machines, which in effect would amass into the first wind-electric generating station. Priorities changed within Russia and the exigencies of World War II altered these plans. The solitary Russian 100kW machine ran for ten years, but when it failed during the war years engineers dismantled it.[4]

Palmer Putnam followed the Russian experiment with particular interest. Practical visionaries are a rarity. Usually, a person is either practical or a visionary, but seldom do such qualities mesh. In the case of Putnam, however, they did. He came from a publishing background, and was president and chairman of the board of G. P. Putnam's, Sons, from 1930 to 1932. However, Palmer Putnam was more devoted to invention and engineering, and he demonstrated skill in both.

Putnam's interest in wind energy originated in a pragmatic way. In 1934 he built a house on Cape Cod and, as he put it, "found both the winds and the electric rates surprisingly high."[5] He thought of wind energy, intending to have the utility company both maintain standby service and also purchase from him surplus wind-generated electricity. However, no mechanism existed that would convert DC current to AC, facilitating compatibility between the two systems. Putnam researched the problem and soon the economies of scale captured his engineer's imagination. He concluded that "if an economically attractive solution to the problem existed, it lay in the direct generation of alternating current by a very large, two-bladed, high-speed windmill, feeding into the lines of an existing hydro, or steam and hydro, system.[6] He would demonstrate that such a novel form of energy creation was both economical and practical. Putnam had in mind a machine over two hundred feet in height: it would dwarf both the Brush wind dynamo and the Russian Balaclava experiment. In a 30mph wind, it would generate 1,000 kilowatts (1MW), and in higher winds, 1,500 kilowatts (1.5MW). Except for the Russian experiment, nothing approaching this scale existed. A successful experiment would require basic research.

Perhaps, a few decades later, Putnam would have been awarded a grant from a government agency, but he chose to work through private industry. S. Morgan Smith Company, manufacturers of hydraulic turbines, were seeking a way to diversify their product line and wind turbines offered an interesting possibility. The company's directors decided to underwrite the project.

Putnam took primary responsibility for the project, but he also sought the best scientific advice available; he was a team player.

The Smith-Putnam turbines on Grandpa's Knob in the Green Mountains, Vermont, represented the first American effort to build a large turbine to feed an electric grid. Designer and engineer Palmer C. Putnam is in the foreground. (Courtesy Cynthia Guy, Lincoln Center, Mass.)

Vannevar Bush, then dean of engineering at the Massachusetts Institute of Technology (later, during World War II, to be Putnam's boss at the Office of Scientific Research and Development) suggested preliminary studies relating to the economy of scale, and then assisted in the research.[7] Professor Theodore von Karman, of the California Institute of Technology, assisted in the general design and John F. Haines, of Engineering Projects, Dayton, Ohio, conducted wind-tunnel experiments at Stanford University. Professor Joseph S. Newell, of the Massachusetts Institute of Technology (MIT) designed the blades, which were then built by the Edward G. Budd Manufacturing Company. The Woodward Governor Company, of Rockford, Illinois, built the speed-control system and Wellman Engineering and General Electric provided expertise with complex electrical mechanisms. The Budd company designed the tower, which was built by American Bridge.[8] The project represented a successful partnership between university and private industry talent. As Vannevar Bush put it in the introduction to a book Putnam published in 1948, the project was "conceived and carried through free enterprisers who were willing to accept the risks involved in exploring the frontiers of knowledge, in the hope of financial gain."[9] Designed during the depression years, it put idle hands and bright minds to work.

Where would Putnam erect this remarkable wind plant, which had surely outgrown his Cape Cod home site? After a considerable amount of anemometer research—balloons were sent up to monitor air-flow over potential sites and tree growth was noted as indication of wind direction—at a meeting at Akron, Ohio, in June, 1940, Putnam announced the choice of Grandpa's Knob in the Green Mountains of Vermont, twelve miles from the town of Rutland. The site boasted several advantages. Except for hunters in search of deer and other game, people rarely tarried in the forest of the Knob. But, of course, Putnam chose the site primarily for the wind. At 2000-foot elevation, air currents were generally steady and strong (up to 140 mph were registered), but not turbulent. Having chosen a site, Putnam was ready to begin construction.

Foresters denuded the knoll, exposing a rocky base, and the assembly team delivered components of the windmill to the Vermont Marble Company plant in Rutland. Assembly itself began during the winter of 1941. Just getting the materials to the top of the Knob proved a herculean task. Two fifteen-degree grade hairpin turns proved particularly challenging, requiring an extra tractor in front as well as a bulldozer pushing from the rear.[10] Truckers eventually hauled some five hundred tons of parts and materials up the newly constructed, 1.8-mile mountain road.[11] On one of the hairpin turns, the only serious mishap occurred. On a windy, subzero day, the crew inched a forty-ton pintle girder on a flatbed truck up the mountain with the aid of two tractors. On the turn, the girder broke free, rolled off the truck, and landed, upside down, in a snow bank. Fortunately, no one was injured and the girder survived undamaged.[12]

Considering the novelty and the scale of the project, construction foremen had few problems. Not only did they raise the huge tower, assemble the generating equipment, and attach the mammoth blades, they also constructed a so-called bombproof building to house an air-research laboratory and the "electric brain" to control the unit.[13] The bombproof construction did not reflect a security concern, but rather was an engineer's response to the physical environment. Putnam feared that ice would build up on the giant blades in winter—and, with a potential blade-tip speed of 15,785 feet per minute at 28.7 rpm, these blades, as one reporter noted, "may hurl ice chunks with projectile-like speed as far as 900 feet." The building had to be tough in face of such impact potential.[14]

Construction continued through the summer of 1941. By August, the final touches were being made and, on October 19, the Smith/Putnam plant started generating electricity. It operated for approximately sixteen months, producing 298,240kWh in 695 hours of on-line production.[15] On February 20, 1943, a bearing failed at the

A 7.5–ton blade for the Smith-Putnam turbine is trucked to the top of Grandpa's Knob. The construction crew built miles of road that required hairpin turns. The project was completed without serious mishap. (Courtesy Cynthia Guy, Lincoln Center, Mass.)

downwind end of the main shaft. In early 1943, with national priorities focused on the war effort, Putnam and his colleagues went to work at the Office of Scientific Research and Development, designing amphibious landing craft and experimenting with an exotic snow machine for use in covert military operations in Norway. He had little time or energy for his crippled wind turbine.[16]

Twenty-four months passed before workers found and installed a new main bearing and the wind turbine came back on-line. However, within weeks an even more catastrophic event disabled the turbine forever. At 3:10 a.m. on March 26, 1945, during a smooth southwest wind of 20mph, one of the blades separated from the rotor, careening to the ground and on down the side of the mountain. Three months later, Putnam described the event for *Power* magazine:

> The operator on duty aloft heard nothing, but suddenly felt considerable motion and, working his way against this to the control panel, made a quick manual shut-down in about 2 or 3 seconds. Fifteen seconds may have elapsed after the operator first felt motion, before he brought the turbine to rest. During this time the turbine completed six or seven revolutions with only one blade attached.[17]

Three years later, in his book *Power from the Wind*, Putnam described the event in more dramatic terms:

> Harold Perry, who had been the erection foreman, and was a powerful man, was on duty aloft. Suddenly he found himself on his face on the floor, jammed against one wall of the control room. He got to his knees and was straightening up to start for the control panel, when he was again thrown to the floor. He collected himself, got off the floor, hurled his solid 225 pounds over the rotating 24-inch main shaft, reached the controls, and brought the unit to a full stop in about 10 seconds.[18]

Inspection showed that a number of cracks had developed along the blade spar, particularly at the bolted shank/spar connection. Inspection of the one surviving blade also revealed cracks. Putnam presumed that stress and fatigue caused the mishap.

The disastrous collapse did not discourage Putnam. He saw the crash of the blade simply as part of the pitfalls of a new technology. In *Power* magazine, he reassured readers that "we believe we can design and build wind turbines that will be entirely practical. . . . REBUILDING OF THE UNIT IS UNDER WAY."[19]

But rebuilding did not occur—for several reasons. First, the materials were difficult to obtain. At a time when wartime restrictions were still in force, the wind plant had not been designated as a priority. Second, replacing the blades would have been costly. As Vannevar Bush had noted with pride, the Smith/Putnam project had been "conceived and carried through free enterprisers who were willing to accept the risks involved in exploring the frontiers of knowledge, in the hope of financial gain."[20] This was certainly true, and to their credit the board of directors and stockholders of the S. Morgan Smith Company had made an investment of over a million dollars with very little return. The company could not be expected to do more. And third, although wind energy was fascinating, engineers considered it to be irrelevant. Nuclear energy was on the horizon, and both the private sector and government were geared toward investing in what appeared to be the ultimate solution to the question of energy supply.

But the Smith/Putnam wind plant had demonstrated that wind, that exotic power source, could be harnessed on a large scale. Writing in 1948, Vannevar Bush pronounced: "The great wind-turbine on a Vermont mountain proved that men could build a practical machine which would synchronously generate electricity in large quantities by means of windpower."[21] Bush went on to state that windpower could be economical. Ignoring the already existing army of small scale-wind plants, he forecast that "at some future time homes may be illuminated and factories may be powered by this new means."

Whether or not Bush was aware of the thousands of Jacobs and Wincharger machines that lit homes throughout rural America, he could be excused for stating his prediction in these terms: his engineering proclivities lent themselves to the large and the complex. Putnam and his scientific cohorts thought only in terms of grid systems—such as that of the Central Vermont Public Service Corporation. For them, the genius of their wind machine was not only its size but the fact that it fed into an AC central system. This was a significant achievement. In 1942, engineer John Wilbur wrote: "Alternating current is being developed from wind energy for the first time."[22] Putting it in layman's terms, Wilbur explained that "this requires careful speed regulation, since it becomes necessary to drive the AC generator, which in this instance is synchronous, at a constant speed, while wind conditions are varying, in order that the frequency of the system to which the wind turbine is connected may be matched." Putnam and his team collected a disparate, chaotic energy source and integrated it into a steady, orderly system—an engineering challenge that required skill and patience. The Central Vermont Public Service Corporation executives had recognized that wind power would be most efficient when used in concert with a hydroelectric system and Wilbur explained that in such a case, when the wind blew, the hydroelectric units could be shut, retaining the water in storage; when the wind ceased, the water could be released to generate power. Such a system raised total power output significantly. "In effect," wrote Wilbur, "when the wind is blowing, wind energy can be stored by means of the water withheld in the reservoirs."[23] This symbiotic relationship of the two primary kinetic energy sources—wind and water—has been often recognized since the 1940s and has been employed successfully in more recent projects.

Although the Smith/Putnam turbine succeeded as an experiment, it failed in terms of economy of scale. The expense of repairs in the event of a breakdown proved to be prohibitive. The task of replacing and repairing machinery 120 feet above terra firma was painfully expensive—if parts and labor were available at all. Putnam, as Vannevar Bush noted, built the wind plant in the hope of profit. Breakdowns eliminated this hope. The expense of putting the plant back into operation would have been in the neighborhood of $300,000, an outlay that the Smith company could not justify.[24] A smaller plant could reduce such costs: economy of scale might be realized through multiple, medium-sized units—not by huge plants such as Grandpa's Knob.

After the March 1945 disaster, the Smith/Putnam machine never again operated. Putnam placed patents in the public domain, then shifted his engineering skills in new directions—primarily toward nuclear power. But his legacy still has an impact. Almost thirty years after the last revolution of his wind turbine, in an era of new interest in alternative energy, engineers had not forgotten that Putnam "has shown us the way." Speaking in 1974, electrical engineer William L. Hughes allowed that Putnam's machine "was really the precursor of all of the wind work that is being done today."[25]

PERCY THOMAS—PUSHING THE WIND

Whether wind power will advance to any great extent in the future, or whether the need will be obviated by nuclear development only time will tell.

—*ENGINEERING JOURNAL*, 1953

That thirty-year period was, in fact, the nadir of modern interest in wind energy. Systematic, large-scale exploitation of the wind was not, however, without its advocates in the post-World War II years. Percy H. Thomas, for one, eagerly followed the Grandpa's Knob experiment from his office with the Federal Power Commission. Thomas, a high-ranking engineer, found an opportunity to jump into wind energy research when the Central Vermont Public

The Smith-Putnam turbine at work at night. The 1.5MW machine performed well for sixteen months, but when a blade spun off, World War II emergency regulations made it impossible to get the mammoth generator repaired. Much knowledge was nevertheless gained from this pioneer engineering effort. (Courtesy Cynthia Guy, Lincoln Center, Mass.)

Service Company listed the Smith/Putnam wind turbine as one of its operating units. Since Congress charged the Federal Power Commission (FPC) with studying the production of electric power—in whatever way produced—Thomas was thus given a convenient entrée to continue a fascination he had felt since his college years at MIT. With the support of his boss (William Warne), he began what turned out to be eight years of research—from approximately 1943 to 1951. Designing and building a major wind plant became an elusive life goal for Thomas. Later, he would testify that the opportunity to engineer something as novel as a huge wind-energy plant "happens once in a generation, and to be allowed to work on it is a very great pleasure."[26]

Thomas did not participate directly in the Grandpa's Knob experiment, but he followed it so closely that he can be considered almost part of the team. He believed in wind power and the concept of renewable energy; although a public servant, he profited by the activity of private industry. One might expect Thomas to have recognized that the economy of scale, with regard to repair costs, did not work to the advantage of the massive Smith/Putnam unit, but Thomas believed that the problem lay with the blades. Engineers who drew up the plans mistakenly provided incorrect figures to the contractor. When the engineers realized the error, they could not afford the price tag ($100,000 to $150,000) to correct it. Furthermore, according to Thomas, because of the prohibitive cost, the Smith/Putnam blades were not of the optimum shape.[27] Through his association with private industry, and benefitting from the knowledge, experience, and past errors of Putnam and his staff, Thomas felt confident that he could build a trouble-free machine.

The wind plants he proposed were unprecedented in both power and size. In the design of a 7,500-kilowatt plant (7.5MW), his largest, Thomas pushed the economy of scale to the limit. His design called for a 575-foot-high wind turbine, the blades reaching higher than the Washington Monument.[28] Each of the six V-shaped legs of the tower would be approximately 280 feet across at its base, tapering upward in a manner somewhat resembling the Eiffel Tower.[29] Besides being of extraordinary size, the Thomas model was unique for its twin nacelles, each featuring seventy-five-foot blades. Twin plants, suspended from a single tower, had never been designed before, nor has such a design been used since Thomas's efforts. But to this visionary engineer, it made perfect sense. If an engineer designed an expensive and huge tower, why not maximize the available wind, using two rotors rather than one? It was through this twin-rotor design and improved blades that Thomas expected to build to higher efficiency. He planned on a 7.5-megawatt output in a 34mph wind.[30]

Thomas and the Federal Power Commission administrators envisioned that such immense wind plants would be used in conjunction with hydroelectric production. In an illustration, Thomas showed three 7.5-megawatt plants, placed along a ridge, producing energy, with power lines stretching toward an industrial city. In the distance, power lines flowed from a reservoir and a hydropower house. Like Putnam and others, Thomas envisioned that his wind machines would operate in conjunction with the hydroelectric system. In Thomas's words:

> In this case, the hydro-storage reservoirs and the other reserves provide for firming the hydrogeneration during dry years, but they are suitable also to firm wind energy, the storage thus serving double duty. This would seem to limit the capability of the storage for its hydro duties but since the deficiencies of the wind are limited to a few days while the water lack in the rivers is for months or even for a year, the use of the storage for firming the wind handicaps it very little in its original work.[31]

Thomas excelled at designing, not explaining. His idea was that the two power sources could work well together, particularly in the West, where water was available but not abundant and wind was abundant but not necessarily available. With wind-energy development, water in reservoirs could be held until such time as the wind ceased to blow.

In 1950—seven years into his research—Thomas and his superiors at FPC and the Department of the Interior prepared to take his ideas and plans before Congress, hoping to secure federal monies to build a demonstration wind generator. House bill 4286 (82nd Cong., 1st Sess.) would provide over $2 million to complete research, locate a site, and construct a giant plant along the lines Thomas had proposed. Although a private company might act as the principal contractor, the power generated would be integrated into a federal project or system. The bill justified the experiment in order "to promote the conservation of the natural resources of the Nation and to aid in the national defense." The wind machine—not to exceed 10,000-kilowatt (10 MW) capacity—would also determine and demonstrate the feasibility of integrating wind-produced electricity with that of more traditional and commercial energy sources. Other arguments for the development of wind-energy systems included the fact that the "fuel" required no labor or transportation costs, that the units necessitated only a short construction schedule or lead time, and that the material requirements of such plants were minimal, particularly because no storage facilities would be involved.[32]

The references to conservation and national defense deserve further comment. The notion that the nation might exhaust its nonrenewable energy sources was not new, but the idea that companies might *preserve* those nonrenewable resources through the development of renewable sources was novel. When assistant secretary of the interior William Warne explained to congressmen that wind-energy plants "might save oil that otherwise would be used as fuel in steam plants," he was not advocating an altogether new concept, but it certainly had a fresh slant. Petroleum conservation had a long history and in 1951 almost every American knew first-hand of that policy through wartime gas rationing. But the possibility of substituting a renewable source of energy to preserve petroleum was relatively unheard of.[33]

Noticeably, no government official or other advocate at the hearing played the pollution card. No one noted that by producing energy through wind power rather than oil or coal, a significant byproduct would be clear air. By the 1950s, air quality was deteriorating in such metropolitan areas as Los Angeles, but officials placed the blame solely on the automobile. It was too early for Americans to grasp the complexity of air pollution, or comprehend the difference between a clean plant and one that polluted.

A more chilling concern for Americans at that time was national security. By 1949, the Soviet Union had detonated an atomic bomb, years ahead of the predictions made by U.S. security experts. Senator Joe McCarthy roamed the halls of the Capitol and cold war paranoia seized the nation. A large number of citizens seemed prepared to believe him, and to subscribe to the reality of some outlandish plots—such as the sabotage of the nation's power centers. Decentralization of defense plants had proved a popular idea. Since government officials encouraged the placement of industrial plants in the countryside, should not the same apply to the energy source necessary to power such plants? Wind energy had the great advantage of lending itself to decentralism, and assistant secretary William Warne, of Interior, noted the fact in his testimony. The aerogenerators "would be located in primarily isolated locations," and the widely-dispersed, multiple plants would render the system "less vulnerable to damage from attack."[34] Decentralization of power production, so long discarded as unprogressive, had finally found an argument in its favor—capitalizing on cold war fears.

Warne also paraded national pride and an uncertain future in his effort to convince the hearing of the value of the project. He reminded the congressmen that although the nation had a wealth of water and fuel resources, "we cannot afford to let other nations out do [*sic*] us in wind-power development." In other words, the United States must retain the technological edge; when the nation looked "toward the expansion of its power resources," it would thus be ready, and not at square one.[35]

To some of the members of the Committee on Interior and Insular Affairs, the expenditure of monies for such a novel idea proved daunting. Both Warne and Thomas assured the committee that although the experimental plant might cost $2.5 million or more, subsequent units "should be available to both private and public systems at a cost comparable to other types of generating equipment." They believed that, after the initial cost of the prototype, "wind-power plants like it can be designed and built which will be competitive with both steam and hydro power."[36]

Asked about the possible location of an experimental unit, Warne and Thomas confessed that no site had yet been chosen. They had not conducted sufficient research to name the best location. But the site would be in the West, because it was to be integrated with a hydroelectric system operated by the Bureau of Reclamation. One possible site was noted: the Sherman Hill area of southeast Wyoming, near Laramie. This site offered sufficient wind and could be operated in tandem with bureau hydropower systems and nearby transmission lines.[37]

The committee received Thomas and Warne warmly. Curmudgeon Wayne Aspinall, of Colorado, said it was his opinion that the bill would not pass, but other committee members seemed supportive. Some members found themselves unable to ask substantive questions about such a novel idea. Hamer Budge, of Idaho, stated simply: "This would certainly be a wonderful thing if it works." William Yorty, of California, was impressed. He found the possibilities "so attractive I do not see how we can afford not to use them." Congressman Fine, of New York, said he was "so intrigued" by Warne that he skipped another subcommittee meeting and congresswoman Bosone, of Utah, bordered on the fulsome in praise of Thomas and his ideas. She believed him to be the archetype of the engineer with "honesty, humility, and scientific brain, and the patriotic pursuits of the scientific idea." She considered him "the greatest asset for the whole idea."[38]

But Aspinall's prediction of failure proved accurate: the bill failed to survive the pitfalls and gauntlets of the legislative process and a similar bill was not presented later. Thomas's engineering designs went untried and in the United States the development of knowledge about wind energy came to a standstill. In Denmark, France, and England, a little experimental work continued.

One can point to a number of reasons for the defeat. The failure of the Grandpa's Knob frequently provided the focus of discussion in the committee. The legislators considered Putnam's blade failure to have proved the whole endeavor a washout. The engineers, Thomas, and Warne disagreed with this view. Warne maintaining that the installation "did operate, and it operated on a commercial basis."[39] Another factor was that the bill suffered from lack of a local advocate. Because the Department of the Interior had not chosen a site, no legislator, in either the House of the Senate, emerged as advocate. Bluntly put, the bill was no one's pork barrel.

Even given a different scenario—if the Smith-Putnam turbine had performed perfectly and the political ducks had been all in a row—funding of the Thomas project would still have been unlikely. The defeat of wind power in 1951 had more to do with outside forces and interests than with performance or potential. There was a hidden agenda: the international political climate and nuclear power. These nullified serious interest in wind energy. Shortly after the United States military dropped atomic bombs on Hiroshima and Nagasaki, bringing World War II to a conclusion, both scientists and politicians attempted to move the focus of atomic energy in a different direction. For understandable reasons, leaders preferred to stress not the moral uncertainties of the bomb but the benefits to people that this monumental new energy source could provide. By Christmas 1945, Senator Brian McMahon had introduced a bill that called for such a new direction—one away from the carnage of war. The

Congressional Declaration of Policy of the McMahon bill stated that, although the effect of civilian use of atomic energy could not be determined, "it is reasonable to anticipate . . . that tapping this new source of energy will cause profound changes in our present way of life."[40] By late 1946 the bill became the McMahon Act, a legislative achievement that would turn the nuclear effort toward "improving the public welfare, increasing the standard of living, strengthening free competition among private enterprises so far as practicable, and cementing world peace."[41]

Americans now anticipated a nuclear age, and, surely, bountiful energy would be its hallmark. As David Lilienthal, former Tennessee Valley Authority head (and, in late 1946, the first chief of the AEC—the Atomic Energy Commission) stated: "Reasonable men spoke of atomic power so cheap it wouldn't pay to meter it."[42] Euphoric expectations surrounded the peacetime possibilities of nuclear power. This new development might effect U.S. society as profoundly as had the electric light, the telephone, the automobile, and the airplane *combined*. Scientists had developed something remarkable: a new *primary* energy source—one that seemed to hold unlimited possibilities.

Certainly most members of Congress embraced this promise and they provided lavish funding for the Atomic Energy Commission. Lilienthal noted: "Never had there been, anywhere in the world, so broad and generously financed an attack on the technical and economic problems of bringing a new scientific discovery into the realm of application and of widespread use."[43] A later critic of nuclear energy development noted that by 1953 the AEC bureaucracy was so vast that it "owned and operated three towns; it employed sixty-five thousand construction workers, five percent of the nation's entire construction labor force; it consumed ten percent of the country's electric power." By the close of that year, the AEC's capital investment topped "nine billion dollars, more than the combined 1953 investment of General Motors, US Steel, Du Pont, Bethlehem Steel, Alcoa and Goodyear."[44]

Was all this incredible cost worth it? Perhaps it was, from a military or cold war perspective, but not for electricity production. By 1963, Lilienthal had changed his mind. He distanced himself from nuclear power, stating that scientists had not demonstrated a way to dispose of the "furiously radioactive" byproducts. Furthermore, not a shred of evidence existed to indicate that nuclear power was the least expensive method of power production. Even if it was, the dangers involved if there were an accident were horrendous—too horrendous to risk. Lilienthal favored a less glamorous power source, such as coal.[45]

Lilienthal's views were prescient, but he did not represent mainstream thinking. Congress was convinced by scientific experts that peaceful use of atomic power would transform society, and nuclear power plants would usher in that new society. A nuclear power plant here and there would obviate the need for oil or coal plants (and certainly for wind machines). Thus, Congress appropriated billions of dollars for research and development of peaceful uses of nuclear energy, and not a cent for alternative forms of power production. The good scientific work of Palmer Putnam went for naught: scientists and engineers discarded that knowledge in their obsession with nuclear power. Congress chose wrong; the taxpayers lost, and so did alternative energy development.

PART TWO

MODERN WIND-ELECTRIC

CHAPTER 7

IN THE LEE OF THE ENERGY CRISIS, WIND PICKS UP

Inventors and technicians will have to pay homage to windmills. They'll have to build machines that use, not abuse, the unearned gifts of nature.

—STUART UDALL, FORMER SECRETARY OF THE INTERIOR, SPEAKING IN 1971

IN THE LATE 1950S AND EARLY 1960S, THE UNITED STATES BECAME ELECTRIFIED. All-electric homes became commonplace, as did office buildings featuring windows that did not open and lights that did not have light switches. The lights stayed on at all times.[1] Energy seemed limitless and Americans who, during the war had conserved, afterwards consumed. The Edison Electric Institute, the utility companies, and other industry promoters launched a campaign for consumption that proved so successful that in the two decades after World War II Americans increased their electric power usage by 7.8 percent each year, thus doubling consumption in ten years. It was what one writer called "a golden age for electric companies."[2]

With consumption progressing at a splendid rate, the industry faced a predicament. The seemingly ever-rising growth curve required expanded energy production: How would the demand be met? Both investor-owned companies and public power companies might have looked to fresh possibilities; instead, they retreated to the dependable "hard energy path"—that is, reliance on coal, oil, natural gas, and, in trivial amounts, nuclear.[3]

In Europe, however, experimentation with wind energy continued, particularly in France and Denmark. Denmark proved particularly successful with the 200kW Gedser turbine, using a 78-foot-diameter rotor. The Gedser proved both efficient and reliable, running continuously for ten years (1957–1967).[4]

But the Danish turbine created nary a ripple of interest in the United States. Immediate demand would be met through new fossil-fuel plants. As for the future, most utility experts assumed that emerging nuclear technology would fill any gaps, providing an energy cornucopia to meet the most inflated energy forecast. Nuclear scientists were particularly enthusiastic, and not only for technological reasons. As a penetrating book put it: "For them, the transition to civilian atoms represented an indispensable social legitimation of their scientific interests, the irrefutable historic justification of the primacy of nuclear physics in the field of science and technology."[5] Thus, the idea of redemption (or guilt) for the destructive power of the atomic bomb played a part in the scientific community's dedication to nuclear power.

These scientists found powerful allies. Perhaps at no time in U.S. history had the national leadership been so mesmerized by the potential of technological progress. This infatuation was not without cost to the taxpayer. Between June 1955, and June 1964, the federal government, primarily through the Atomic Energy Commission (AEC), invested $27 billion in the development of nuclear energy. During the 1950s, private capital contributed only 12 percent toward the government-sponsored nuclear reactor program. Commercial nuclear power received a subsidy unparalleled in the U.S. experience.[6]

In spite of the AEC's bureaucratic enthusiasm, utility companies were hesitant. With reason, they feared this awesome new power source. The potential financial liability of a China Syndrome (meaning an overheating and meltdown of uranium fuel rods) struck dread into the hearts and heads of utility executives and their insurance companies. In spite of the AEC's enormous financial support and constant encouragement, the utility companies refused to budge until 1963.[7] In 1963 and 1964, Westinghouse and General Electric offered turnkey nuclear plants at prices that were most attractive when compared with coal-fired plants. The two companies lost money on their bids, but nuclear reactor orders jumped from seven in 1965 to twenty in 1966 and to thirty in 1967.

That decade (1965 to 1975) represented a brief halcyon era for nuclear power. The environmental movement and the antinuclear forces had gained strength, but protesters were unable to slow the nuclear juggernaut once it was in motion. Nuclear reached its high point in the early 1970s, when utility companies ordered an unprecedented number of nuclear reactors, partially in response to the 1973 oil crisis. Needless to say, it was a decade in which little thought was given to wind energy. Virtually no money was spent on wind research and development.

Quite suddenly, the infatuation with nuclear plants came to a halt. Skyrocketing costs, increasing antinuclear protests, and indecision regarding nuclear waste all contributed to the turnaround. Utility companies had ordered 231 nuclear plants through 1974; after that, only 15. Since 1978, no plants have been completed. Most orders were canceled, and dozens of partially completed plants have been abandoned.[8] The Three Mile Island incident (1979), in which many of the vaunted safety features of a nuclear plant failed, causing radiation leakage and a near meltdown, shattered public confidence. Some years later, nuclear energy suffered another calamity when the nuclear plant at Chernobyl, in the Ukraine, exploded, spreading radiation damage and death over a vast area.

More trouble came on the financial front. When the five-plant nuclear project of the Washington Public Power Supply System (WPPSS) declared bankruptcy and defaulted on $2.25 billion dollars worth of bonds in July 1983, the struggling industry became moribund. One critic remarked that WPPSS "promised power without cost, and they delivered cost without power."[9] It is ironic that recently some of the twenty-three utility companies that comprised WPPSS have considered wind power to meet their additional needs.

In other countries, nuclear was running into similar problems[10] and obstacles facing the development of large wind machines seem minuscule alongside the financial fiasco of nuclear development. One U.S. commentator noted ironically that

> records for [the] costliest single [nuclear] power plant cancellation are epic but transient; Washington State's 1982 mark of $2.3 billion gave way in the first month of 1984 to a still more robust $2.5 billion collapse in Indiana. Either New York's Shoreham unit or New Hampshire's Seabrook Station may seize the lead later this year [1984], by abandoning the field after expenditures exceeding $3 billion."[11]

At no time in U.S. history has science and technology, in concert with federal officials, promised so much and produced so little. It represented a major defeat for U.S. technology, and one that cost not only ratepayers but taxpayers, too, billions of dollars. Whether the nuclear power industry can be revived is questionable. Pro-nuclear advocates such as Bernard Cohen argue for the efficiency and safety of the nuclear option, but it is unlikely that the U.S. public will be convinced in the near future.[12]

The demise of nuclear power assisted in a resurrection of interest in wind energy, but it was not the only factor: other significant social concerns were coming together. The environmental movement, with origins in the early 1960s, came to a crescendo with Earth Day 1970.[13] Significantly, the movement considered *alternatives*: alternative lifestyles and alternative energy sources. The new national mood questioned business as usual—and that meant questioning the construction of more huge power plants. Nuclear was unsafe and costly: fossil fuels consumed nonrenewable resources and polluted the air.

By 1970, pollution issues no longer concerned just a few environmentalists. Gallup polls surveying U.S. public opinion on major domestic problems in 1965 and 1970 indicated that the problem of "reducing pollution of air and water" had moved from a ranking of ninth to second place, exceeded only by "reducing crime."[14] Congress and the president responded with the passage of the Clean Air Act of 1970, placing more pressure on utility companies to reduce production of carbon dioxide (a principal contributor to global warming), nitrogen oxides (a cause of acid rain and air pollution), and polluting smoke and particulates. An energy source—such as solar—that was safe and left the environment *unchanged* seemed particularly attractive. Here and there, prominent leaders spoke of wind energy. Stuart Udall, who had been secretary of the interior under President John Kennedy, called wind generators "symbols of sanity in a world that is increasingly hooked on machines with an inordinate hunger for fuel and a prodigious capacity to pollute." Udall praised wind generators as devices that "respect the laws and limits of nature." He reasoned that if Americans were to meet the challenge of the future, "inventors and technicians will have to pay homage to windmills. They'll have to build machines that use, not abuse, the unearned gifts of nature."[15]

However, although the audience for such views was increasing, only a minority of Americans thought in such terms. Many utility companies, seeking a prompt solution, chose the so-called hard energy path, switching from coal to oil. This move turned out to be only a Band-Aid for a seriously sick industry. Americans, becoming increasingly aware of the health hazards of electricity produced by oil or coal, wanted change. In 1973, the U.S. Senate held hearings on a bill aimed at providing research funds to produce "electric energy with minimum impact upon the environment."

In opening the hearings, Senator Frank Moss, of Utah, voiced the feeling of millions of Americans:

> For many years such work [producing energy with minimum impact on the environment] was neglected by both the Federal Government and the industry. Today we are paying the price of that neglect: The Nation is chronically faced with the threat of power brownouts and blackouts, citizens oppose the construction of electric facilities, fossil-fuel electric powerplants rank with the automobile as the Nation's worst polluter, and our energy systems are extremely inefficient—two-thirds of the fuel consumed to generate electricity is discarded as waste heat.[16]

But no matter how logical or necessary Senator Moss' appeal, such views were not the key for most Americans. It was inflated oil prices and gasoline shortages at the pumps that forced people to reevaluate. In 1973, the cost of oil skyrocketed, a consequence of the Organization of Petroleum Exporting Countries (OPEC) October oil embargo. This embargo had deep roots. Factors involved included the multinational energy companies, a valuable natural resource that was priced too cheaply, and the continual political turmoil of the Middle East.[17] These roots had, in 1959, tapped a reservoir of Arab bitterness over colonialism and exploitation and resulted in the formation of OPEC. Throughout the 1960s, the OPEC nations continued to be pawns in an international game of chess over oil, but by 1971 the pawns occupied certain strategic advantages. In the fall of 1973, they called checkmate. A combination of dramatic price increases and the October 19, 1973, announcement by King Faisal of Saudi Arabia of an oil embargo on supplies to the United States threw what had been a fairly orderly international energy system into chaos.[18] Within two months, the price of oil had quadrupled. Long lines of automobiles snaked around any U.S. gas station that had supplies. The gluttonous American appetite for oil found the cupboard bare. The decades of cheap oil were over.

The oil embargo ended on March 18, 1974, six months after it had begun.[19] Within that short period, significant changes occurred. Energy conservation became part of the American lexicon. The Detroit automobile industry commenced designing more fuel-efficient vehicles. Americans realized the vulnerability of the nation to foreign influence. The government looked for new sources of oil—and new sources of energy.

THE SUN COMES UP: FUNDS AT LAST

Scientists and engineers now returned to some old concepts: they resurrected wind energy, that relic of the past. Engineers would clothe this ancient, natural energy source in the garb of computers, high-tech materials, and the language of the space age. It would, however, still represent a basic, ancient idea: Harness the wind for the service of man.

A new energy prophet proposed to capture the wind on a grand scale. Engineering professor William E. Heronemus had plenty of experience with energy. As a U.S. Navy captain, he assisted in the construction of the first nuclear submarine, the USS *Nautilus*. When he tired of working for Admiral Hyman Rickover and the navy, he accepted an engineering position at the MIT. Heronemus admired the work of Palmer Putnam and Percy Thomas—the two pioneers whose work is described earlier—but he had no burning desire to enter the world of renewable energy. However, becoming aware of the pollution of the Connecticut River and the threat of a nearby nuclear plant, he changed the direction of his research. Heronemus became an environmentalist, embracing the notion of clean, sustainable energy sources. "Need we bother very much at all with fission and fusion if there are alternatives?" he asked colleagues in the American Society of Mechanical Engineers and the Institute of Electrical and Electronics Engineers in 1972. "Would there be some unacceptable stigma to our society if we were to opt for an energy system whose science and technology would be very unsophisticated?"[20]

Heronemus described wind energy as an "unsophisticated technology," but his plans were grandiose. Decrying the views of people such as Chauncey Starr, who argued that little significant potential for wind energy existed in the United States, Heronemus forecast that there would be an array of wind turbine stations (each with twenty two bladed turbines, fifty feet in diameter, distributed on a 600-foot tower) in each square mile of a proscribed Great Plains area. This would be equal to "189,000 Megawatts of nuclear central plant installed capacity." Such a contribution, he advanced sarcastically, "might properly be called significant."[21]

Presenting his views at a meeting in Masachusetts, Heronemus discussed the local potential for wind energy: "At the expense of some visual pollution, New England could obtain about 20 percent of her projected 1970 to 2000 generating plant growth in this way" This, he said, was in addition to developing storage of wind-generated electricity, particularly through conversion to hydrogen. Heronemus concluded that "there is a strong case for revived interest in Wind Power. It could be very competitive. . . . It certainly can be the essential ingredient of pollution-free power systems. And it is such a gentle alternative to high temperature combustion, fission and fusion schemes!"[22]

Coming in 1972, such rhetoric fell on sympathetic ears. As noted above, critics of every stripe were challenging the supposed benefits of a nuclear future and the utility of fossil fuels. Perhaps the environmental benefits of this ancient technology were worth the risk of a little federal research and development money? Well over $50 billion had been lavished on nuclear energy, but not a farthing had gone into testing the notion of creating energy from a renewable resource, the wind, for example. Thomas had made an eloquent but unsuccessful plea in 1951; now the calls for aid fell on more sympathetic ears.

In 1974, alternative energy sources, primarily solar, finally had their day—albeit a cloudy day. President Richard Nixon, facing a growing energy crisis, fell back on coal and nuclear for both short-range and long-range solutions. In a Message from the President of the United States Concerning Energy Resources, delivered April 18, 1973, Nixon predicted that nuclear energy would provide 25 percent of electrical power by 1985 and over 50 percent in the year 2000.[23] In his 1974 budget, 73 percent of the research and development funds were allocated to nuclear power. But, in spite of this continuing compulsion to rely on nuclear as savior, Nixon did allow that "solar energy holds great promise as a potentially limitless source of clean energy." He continued: "My new budget triples our solar energy research and development effort to a level of $12 million."[24] Critics underscored that this was a questionable commitment. The funds allocated to renewable energy were paltry; but a tiny piece of the pie was progress, compared with the old starvation policy.

The appropriation swelled immensely with the passage of the Solar Energy Research Act of 1974. Previously the National Science Foundation (NSF) coordinated the limited solar research and development effort. The new act established the Solar Energy Research Institute (SERI) to oversee the effort. SERI, located within the Federal Energy Research and Development Administration (ERDA), was funded relatively lavishly—$600 million spread over a five-year period. Congress empowered SERI to utilize the technical expertise of such agencies as NSF, the National Aeronautics and Space Administration (NASA), the Department of Agriculture, and other agencies. It could capitalize on the resources available through government science and engineering laboratory centers such as Los Alamos and, particularly, the Lewis Research Center, Cleveland, administered by NASA.[25] And it could negotiate outside contracts, employing the expertise of private firms, scientific organizations, and educational institutions.

Under this new legislation, the federal wind-power program raced ahead, consuming $380 million between 1973 and 1988.[26] From the outset of this era, large wind turbine development captured the imagination of U.S. engineers and administrators. Over the years, SERI expended from 75 to 80 percent of available funds on developing a wind turbine of from 1MW to 3MW. If successful, it would demonstrate the benefits of economies of scale. Louis Divone, head of the Wind Energy Technology Division of the Department of Energy (DOE), maintained that "costs should come down as the size of the wind machines goes up," and his team set out to prove that assumption.[27] When Representative Richard L. Ottinger, of New York, asked William H. Robbins, manager of the Wind Energy Project Office at NASA's Lewis Center, if the economies of scale really worked, Robbin was quick to respond: "There is no question in our minds that between intermediate and large machines there are definite economies of scale." He was even specific: "We believe that 200- to 300-foot in diameter machines will produce electricity at the lower cost."[28]

From SERI and the Lewis Center money filtered out for experimentation and prototypes. Alcoa Corporation teamed with Sandia Laboratories to produce a vertical-axis, Darrieus-type 100kW turbine. Prototypes of this model were set up at Rocky Flats, Denver, and at Bushland, Texas, under the direction of R. Nolan Clark of the Department of Agriculture. Other major companies that engaged in the search for an efficient, reliable multimegawatt wind turbine included Boeing, McDonnell Douglas, Hamilton-Standard, Grumman Aerospace, General Electric, and Westinghouse. All these contractors had the scientific expertise to design and build large wind turbines.

These companies and the NASA engineers at the Lewis Research Center directed their primary effort to the MOD (Modification) program. MOD was devised to build and test a number of large wind turbines. The Lewis Research Center designed the first turbine, designated MOD O; a 100kW unit erected at NASA's Plum Brook site near Sandusky, Ohio. It became operational in the autumn of 1975, but quickly failed. Engineers called the next prototype the MOD OA, a 200kW turbine. The Westinghouse Electric Company built four prototypes, erecting them at Clayton, New Mexico, the island of Culebra, Puerto Rico, Block Island in Rhode Island, and Oahu, Hawaii. The next effort resulted in the MOD 1, a 2MW (2,000kW) machine, erected by General Electric on a hilltop near Boone, North Carolina.

These were so-called first-generation turbines. Researchers designated them as research tools, designed to help them gain knowledge of "operational loads, environments, and . . . different types of utility networks."[29] In other words, whether or not the early MOD machines worked would not determine the success of the effort. Of course, these machines were highly visible to the public—a public that did not always understand that inoperative machines could still provide knowledge. Rather, many people saw them as a symbol of the unworkability of wind power.

In 1976, the Boeing Corporation entered the field with the first second-generation machine, the 2.5 MW MOD 2. The MOD 2 turbine featured an upwind, two-bladed, teetered rotor with pitchable tips. The teetered rotor consisted of a hinged connection of the blades to the hub. This teetered hinge allowed five degrees of teter relative to the normal plane of rotation. When working properly, this reduced blade loads and the vibration inherent in two-bladed rotors. The pitchable tips controlled both power in high winds and the start and stop cycles. An advanced "soft" gear train also reduced loads and vibration. Moreover, engineers had decreased the weight dramatically.

The big news was that NASA executives considered the MOD 2 to be commercially viable, "designed for quantity production." They predicted a cost-of-energy of less than four cents per kWh when the turbine was located at sites of 14 mph (annual average)

This huge Hamilton-Standard 4MW unit on the Wyoming plains came out of government interest in alternative renewable energy in the early 1970s. It was funded in 1982. The unit never worked well and in January 1994 a storm destroyed the turbine, ending the experiment. (Author photo)

wind speed.[30] Three of these turbines were erected on the Goodnoe Hills, near Goldendale, eastern Washington state; one appeared in California; another was erected at Medicine Bow, Wyoming.

After MOD 2 came a huge machine from Hamilton Standard (a subsidiary of United Technologies Corporation). The WST-4, a 4MW turbine, had a nacelle "the size of a railroad box car [placed] on top of a hollow-steel tower 80 meters above the ground." The WST-4 was erected at Medicine Bow, Wyoming.[31]

The most recent, and probably the final, DOE and NASA project featured the MOD 5B, a 3.2MW turbine designed by Boeing.

Erected at Kahuku, Hawaii, in early 1987, it was announced as the largest wind turbine ever constructed.[32] The DOE considered all of the turbines to be experimental prototypes; however, the department anticipated that the MOD 2 and the MOD 5B would eventually go into commercial production, with a dramatic lowering in cost per unit and, ultimately, providing inexpensive electrical energy via multimegawatt stations.[33] This hope went unrealized, and chapter 8 offers an analysis of this effort.

THE LITTLE GUYS

While the DOE allocated 80 percent of its wind-energy related funding to megawatt turbines, in the late 1970s it also encouraged the manufacture of smaller turbines, sized from 2 to 40 kilowatts. On the outskirts of Denver, the Rocky Flats testing site—managed by Rockwell International—was host to manufacturers of small wind turbines. Turbines were tested with the objective of perfecting units for both domestic and foreign markets.[34]

This SERI project, it was hoped, would provide the "seed money and technological support" to create a new industry. Assessing it in the 1990s, one could have looked for a better result. Only one or two of the turbines tested at Rocky Flats have had any impact on today's wind-energy scene. There has been successful technology, but it came from a different direction—one that stressed strength and simplicity.

While NASA engineers were designing huge turbines, another group was working on a different scale. In some ways this group was out of step with the engineers, or marched to a different drummer. Some were environmentalists, but to characterize them as such is to oversimplify. Some were members of communes. Some simply wanted to go back to the land. Others had no interest in agriculture but wanted to reverse the direction of American technology. Many were idealists, and even more were antimodernists who wished to simplify their lives. Some shuddered at the nation's growing dependency on foreign oil or dirty coal. Others feared the government and increasing centralization, espousing self-reliance. In Minnesota, the once-revered REA suffered sabotage of its electrical towers from farmer-organized "bolt weevils," rebels so determined to oppose power transmission and centralization that they were willing to face prison sentences for civil disobedience.[35] Others emerged from the antinuclear fight, ready now to embrace renewables. Many recognized that they had a mechanical flare and wanted to work with their hands. They found inspiration not in engineering journals but in the *Mother Earth News*, Stewart Brand's *Whole Earth Catalog*, the Foxfire books, and the words of E. F. Schumacher, the guru of the appropriate technology movement.[36]

Maine farmer Henry M. Clews combined many of these qualities with practical training. In 1973, this twenty-eight year old aeronautical engineer left a teaching position to try his hand at farming. His operation being beyond the service area of the Bangor Hydro-Electric Company, Clews depended on an Australian-manufactured wind generator to keep his fifty-acre farm in power.[37] Clews, like many others, sought technical solutions but disavowed the emerging world of technocracy. One could focus on a number of such "wind-energy environmentalists," and in focusing on two men I hope that they are representative of the many who were caught up in the possibility of actually making a difference.[38] Both Paul Gipe and Ty Cashman were dedicated to moving the energy world in innovative directions—directions that seemed to have been forgotten in the boardrooms of utility companies; to have been forgotten, indeed, by the bulk of the U.S. populace. Gipe had some training in engineering; Cashman had none. Both had climbed plenty of towers, installed and repaired machines as windsmiths, and experienced the dizzying heights associated with such work. Each, in his own fashion, can represent the complex, dispersed, unorganized resurrection of the small wind turbines.

Political activism dominated Paul Gipe's college days in Muncie, Indiana, at Ball State University. He majored in environmental science, but a good deal of his time was spent protesting strip mining and helping to draft a reasonable federal regulation and reclamation bill. President Jerry Ford vetoed the bill, but Gipe felt a sense of satisfaction when President Jimmy Carter resurrected it and signed it into law in 1978. By that time, Gipe was roaming the coal country of South Dakota, Montana, and Wyoming, working for the Yellowstone-Tongue Areawide Planning Organization, a multicounty body dedicated to evaluating "indirect sources" of water pollution for the Environmental Protection Agency (EPA). His task involved locating water-pumping windmills that pierced the aquifer and examining whatever impact there was on the groundwater.[39] The Bureau of Land Management (BLM) had major plans for the Northern Plains—plans that involved mine-to-mouth coal-fired plants (generating plants situated near mines, thus minimizing the transportation of coal). Gipe did his work, but he did not support this scheme to transform the grasslands into generating stations. He had another agenda.

Gipe believed in renewable energy sources and he soon "made a pact to make a difference."[40] One possibility he was noticing all around him. Besides the water-pumping windmills, there were lots of old windchargers—by Jacobs, Wincharger, Parris-Dunn, Sears, Montgomery Wards, Winco. Ranchers had relegated most of the turbines to the junk pile, defering to the high line, but the machines were salvageable for someone with a mechanical flair. With notes he had made while working, letters, telephone calls, and much driving on back roads, Gipe began to acquire wind turbines.[41]

Later, Gipe would describe this period of his life as particularly exciting because his quest to find and resurrect old windchargers was a pacesetter. Eastern Montana, for Gipe, represented a vast scavenger hunt. Often he followed two-track roads or sometimes set off across country in his truck, traversing hills and arroyos.

In the early 1970s, the restoration of old windchargers became a cottage industry. In 1976, Paul Gipe (foreground) and hired hand Dave Ellis pose amid crates of used Montana wind machines often trucking them to Harrisburg, Pennsylvania. (Courtesy John Serbell and Paul Gipe Assoc.)

If luck was with him, he would find a windcharger that could be salvaged. "It was a wonderful experience," he later mused. "Being independent, working in that kind of country and doing something that you felt was useful."[42] Other future leaders within the wind-energy field (e.g., Don Marier and Dan Juhl) also started their careers collecting junked windmills.

When his water-testing job ended, Gipe loaded a tractor trailer with his wind-turbine booty and headed for Pennsylvania. He intended to reconstruct a Jacobs to demonstrate to a local commune that wind energy could provide the answer to energy self-sufficiency. Soon, however, an Erie, Pennsylvania, businessman found him and paid a handsome cash price for all his stock. At that instant, Gipe became "a wholesaler of junk windmills." However, decent wind-chargers were becoming more scarce and the market was limited, so Gipe cultivated another talent—writing. He produced articles for *Solar Age Magazine* and then, in 1983, authored a book on the practical aspects of buying or rebuilding, installing, and maintaining small wind turbines.[43] This resulted in a job offer in California, where he now lives, writing about and working with renewable energy, the dream as intense as ever.

Tyrone Cashman came to wind energy by a different route. Raised in western Wisconsin, he joined the Jesuit order in Oshkosh and was soon sent to study in California, then to St. Louis, Missouri. Cashman thereafter decided that the priesthood was not his calling and, with his theological/philosophical bent, entered the Ph.D. program at Columbia University. When he won his doctorate in 1974, the subject of his dissertation, "Man's Place in Nature," was a harbinger of his future. To the dismay of his German-born professor, Cashman renounced a teaching career and chose a different path. As he later recalled, "By the time I came out with my degree I was tired of words. I had beat them to death."[44] Cashman wanted to approach ecological issues in a practical way and he found his opportunity at the New Alchemy Institute, at Woods Hole, Massachusetts. The institute promoted science that people could use—that is, small-scale technology that did not damage the ecosystem in which it was placed. At Woods Hole, the mission could best be described as promotion of "homestead technology."[45]

Cashman became involved with wind turbines by chance. A position in India beckoned to the institute's wind-turbine expert and, when he left, Cashman inherited the windmills. He found that, unlike many of his colleagues, he did not fear heights—in fact, he enjoyed working high above the ground. He realized the importance of his research following an exchange with Alex Campbell, the premier of the Province of Prince Edward Island. Cashman and others from Woods Hole convinced Campbell that he "could be the engine of the post-industrial revolution" if he promoted wind generation. However, when the premier asked where he could buy a reliable 25kW machine, the New Alchemists replied, "We're sorry, sir, you can't"—and, in a rash, albeit euphoric, moment, they added that the New Alchemy Institute would invent one. A reliable wind turbine eluded Cashman and his coworkers, but, in the attempt at creating one, Cashman gained knowledge and experience.

In 1977, Cashman moved to a Zen Buddhist farm in Marin County, north of San Franscisco, to work with a community interested in conservation projects. In the gulch opening into Muir Beach, he built a colorful, four-bladed sail windmill, based loosely on old Cretan models. In the fall, Sim Van der Ryn, the founder of the Farallones Institute (an organization devoted to the promotion of intermediate technology) took an interest in the windmill and Cashman's work. An architect by training, Van der Ryn held a position at the University of California, Berkeley. He was also state architect in California Governor Jerry Brown's administration and when Brown created an Office of Appropriate Technology, Van der Ryn became its head.[46] Cashman joined the staff as the wind-energy expert and, as events unfolded, played a key role in a legislative effort that was to change the history of wind energy (see chapter 9). Today, Cashman continues to work in the alternative energy field as director of the Solar Energy Institute, located in Mill Valley, California.

THE OLD-TIMERS COME BACK

Meanwhile, people whose expertise in the wind-energy field dated back much farther than that of Gipe or Cashman were feeling a new quickening of interest. Fred Bruns, who had installed many Jacobs machines in the 1930s and 1940s, consulted his old records, recalling customers and installation locations of decades earlier, to turn the new generation on to old machines. Most were still up, and some were operating.[47] Jacobs himself had moved to Florida and changed his manufacturing interests. But with the surge toward new lifestyles and energy paths, he too, had recognized the opportunity to reenter the business. Jacobs did not restore his old turbines—leaving that task for others—nor did he recommence manufacture of the old model; rather, he designed a more powerful (7.5kW) turbine that he felt was more in keeping with the needs of the 1970s. But this time around, Jacobs suffered from the difficulties of economy of scale: his new product did not realized the efficiency or reliability of the old, basic, trustworthy turbines. By the late 1970s, Jacobs retired from the business for a second time, selling his operation and the Jacobs name to Control Data Corporation.[48] The company subsequently evolved into a subsidiary of Keystone Camera Products Corporation, of New Jersey, to be renamed Earth Energy Systems. The Jacobs wind-energy systems are still produced, available through Wind Turbine Industries, of Prior Lake, Minnesota.[49]

To people working in renewable energy, Jacobs became both saint and curmudgeon. On the one hand, he was honored for his pioneering efforts. As Cashman put it: "Marcellus was in business when we were in diapers."[50] Although no longer a manufacturer, Jacobs remained interested in the potential of wind energy. Young idealists admired his accomplishments and hoped to emulate his success; to them, he was a hero. But to electrical and mechanical engineers who focused on the development of huge, two-bladed machines he was a curmudgeon of the first order—an older man who constantly questioned their assumptions and bluntly stated that they didn't know what they were doing. "If they were the ones in the moon program," Jacobs pronounced, "we'd never have gotten off the ground."[51] Furthermore, Jacobs groused that technical knowledge had "regressed since 1960 when we closed down our Minneapolis plant and moved to Florida."[52]

Occasionally, Jacobs did more than carp. In 1951, when Percy Thomas had sought funding for his experimental wind turbine, Jacobs helped kill the bill by writing the chairman of the House Interior and Insular Affairs Committee that they would be well advised to consider a large number of his five-kilowatt machines.[53] Jacobs may have been correct in his analysis, but it was unwise to promote his project at the expense of Percy Thomas's very innovative engineering work. Jacobs's constantly combative style was evident in his 1983 visit to the Altamont Pass wind farm, sponsored by the fledgling California Wind Energy Association. As visitors stood in awe at the sight of hundreds of large wind turbines, a reporter asked Jacobs about his bemused look. "I'm smiling at their mistakes, things I threw away 50 years ago," he said.[54] Jacobs's statement was honest, of course, but the occasion was neither the time nor place for reproach of a struggling industry. In truth, Jacobs did not wear the mantle of idol well. He was too politically conservative for the role. He eschewed the need for, or wisdom of, any government participation in any phase of the industry. Moreover, he opposed large wind generators not only on engineering grounds but for socio-political reasons, too. "The reason for [large turbines] is power companies will own them and sell you the power," said Jacobs. "They don't want you to own them."[55]

This independent streak prevented his participation in the development of a viable trade association, a move that should have been made in the 1930s, rather than the 1970s. When the American Wind Energy Association was formed, in 1974, Marcellus Jacobs soon distanced himself. One founder of the association recalled: "We were eager and waiting to honor him for everything he had done."[56] But Jacobs disdained homage, intent on his separate path as critic. Jacobs possessed mechanical ability and the knack of a salesman, but a certain irascibility limited his participation in the exciting years of wind resurrection. He was committed to competition at a time when the industry could advance only through cooperation.

Jacobs remained active at eighty-three years of age, when an automobile accident, in 1985, abruptly ended his life. There is no denying his huge contribution to wind-power development. He and his brother Joe—along with Charles Brush—must be given accolades in the annals of wind energy history. They invented and built unique machines—machines that worked.

In the 1970s, a number of manufacturers tried to emulate the Jacobs Wind Electric Company's earlier success in manufacturing reliable wind machines. None really succeeded. Small conversion systems, with names such as Northwind, Tumac, and Enertech, made an appearance. Household-name corporations such as Alcoa, Grumman, and McDonnell-Douglas put entries into the wind-energy market. As the manufacturers of the large systems had already learned, success would not come easy. The difficulties they now encountered simply magnified the accomplishment of those early, experimental, self-educated, determined pioneers.

CHAPTER 8

THE RIDDLE OF RELIABILITY

If we had not stopped building them thirty years ago,
we would not have problems today.

—JAMES SCHMIDT, WIND-ENERGY EXPERT, 1981

NEW TECHNOLOGY IS FRAUGHT WITH FAILURE. ONE MIGHT LOOK AT THE early development of electrical systems, or automobiles, to note that for every step forward there were two backward. Executives, corporate and government, often shielded the public from these unsuccessful efforts. Failure occurred at remote test tracks or in the deep recesses of a laboratory, little noticed and therefore causing a minimum of pressure for the industry and the engineers involved.

Some technologies—and their failures—are difficult to mask. Wind energy is one. Like the inventors of the experimental airplanes of old, when each attempt to fly provided good journalistic copy, the wind-energy industry did not hide its blemishes. Wind energy did not suffer anything like the rocketry accidents of the 1960s, but many people noticed that the turbines often did not operate when, it seemed, they should.

Wind energy was particularly visible because wind routes often coincide with highways. Roads and winds have similar requirements in their relationship with the physical environment; for example, winds are strong and reasonably constant in California's low mountain passes. One expert stated that in the U.S. West "all of the mountain passes are viable sites."[1] However, these same passes are traditional trade and transport routes. First, Indians used them, then pioneer wagons, and later came paved highways. In some cases, today, freeways go through the passes. Interstate 10 cuts through the San Gorgonio Pass region: a four-lane road connects the desert with Bakersfield, via Tehachapi Pass, and Interstate 5 winds through the Altamont hills in its descent westward into the San Francisco Bay Area. On each of these major highways, travelers encounter thousands of wind turbines. For many people, assessment of wind energy is based largely on a fleeting observation from an automobile window.

In many cases, the experience has not been positive. At the San Gorgonio Pass wind farm area, the Bureau of Land Management (BLM) and private landowners leased thousands of acres of land skirting freeways to private wind-energy developers. Some of the developers purchased reliable machines; others did not. As one wind-energy developer stated: "A lot of schlocks were getting into the business and selling prototype machines on a mass basis that didn't work."[2] International Dynergy seemed to fit this description. The company operated two wind-electric generating stations in the San Gorgonio area. The one at Maeva park produced only 12 percent of projected kilowatts in the last quarter of 1985; the Cabazon wind park produced virtually nothing. In 1988, Cabazon was described as "an eyesore of broken and twisted blades just off I-10 about ten miles east of Palm Springs."[3] Hundreds of thousands of travelers based their opinions about wind energy on observation of such sites: that the machines were unreliable and the industry offered nothing but a tax dodge for the wealthy.

But it was not just quick-profit artists that made wind energy a questionable venture. In the early 1980s, well-known companies failed miserably in attempts to produce reliable turbines. The failure of the Alcoa 500kW vertical-axis turbine was perhaps the most extraordinary. Placed on the Southern California Edison San Gorgonio site, it began to turn on April 3, 1981. After only two and one-half hours of running time, a computer mix-up caused a brake failure; the rotor accelerated to an overspeed of 250 percent, at which time "one of the blades let loose at the bottom bolts, flew out and hit the guy wires, causing the collapse of the whole torque tube."[4] This catastrophic failure occurred just two days before a wind-energy conference, sponsored by the California Energy Commission, was to meet in Palm Springs. Conference leaders had planned to feature the spanking-new Alcoa and its demise caused considerable embarrassment. Alcoa soon announced that it was leaving the wind-turbine business.[5]

Bendix Corporation tried its hand with a 3MW turbine. The company quickly downgraded the project to 1.2MW when the turbine suffered "unbearable high hydraulic pressure in the fluid drive transmission."[6] Further catastrophy was experienced by the United Technologies Research Center (UTRC). This company, which specialized in helicopter rotors, installed two prototype machines in partnership with the Wisconsin Power and Light Company. In November 1980, both UTRC turbines "suffered a catastrophic failure just ten days after installation." Early in 1982, the company announced the termination of its wind-energy program.[7] *Catastrophic* seemed to be the adjective of choice in describing many of the experiments and prototypes, although the Wisconsin utility company tested other turbines with disheartening—not catastrophic—results. Jacobs, Jay Carter, and Windworks turbines did not suffer total failure. They were, however, subject to "continual stoppages for minor repairs."[8] The commonsense message was clear: Don't invest in a wind turbine.

In an effort to make small wind turbines more reliable, as mentioned in chapter 7 the federal government began its own testing program in 1976. At Rocky Flats, near Denver, the DOE Federal Wind Energy Program (administered by Rockwell International's Energy Systems Group) tested some twenty-five commercially available units. Terry J. Healy, the manager of the testing program, commented that "it is kind of like a rebirth."[9] Healy implied that the knowledge gained would not be a continuation of accumulated knowledge but rather a new start—perhaps like reinventing the wheel. Researchers publicized a few successes, but more frequently they experienced failure. Five years into the program, project leaders made a somewhat upbeat report: "Small wind systems are finally beginning to achieve the potential promised for them."[10] But other assessments, in the same report, were less optimistic. The program involved the installation of forty small systems on the properties of selected users throughout the United States, the units ranging in size from 1.5kW to 40kW. Engineers connected them to local utilities and when thirty-two of the participants responded to a questionnaire, two-thirds of them labeled the performance as "poor."[11]

Clearly, reliability had not been achieved. Critics of federal boosterism argued that the "government has done the public a great disservice by suggesting that wind energy can be used to generate electricity cheaply."[12] James Schmidt, a wind-energy expert, asked rhetorically: "We were using wind generators 40 years ago; how come there are so many problems with using them today?" And he answered: "If we had not stopped building them 30 years ago we would not have problems today."[13] Neither the efforts of private companies nor the infusion of federal dollars could instantaneously overcome the consequences of thirty years of neglect.

ASSESSMENT

The wind was proving to be a stubborn resource to harvest. No company or individual could build a reliable wind turbine suitable for modern uses, especially one able to run year after year with the vagaries of western weather. Andrew Trenka, manager of the wind systems at Rocky Flats, probably spoke for a whole industry when he confessed that "we tended to be blinded because windmills had been used for more than 1,000 years. . . . We thought the technology was there and all we had to do was bring it into the 20th century."[14] The outright failure of some turbines and the questionable reliability of many others, both small and large, could not be hidden from the public. Individual consumers wished to become energy independent and participate in conservation efforts, but the experiences at Wisconsin Power and Light and Rocky Flats suggested caution. Magazines (e.g., *Mother Earth News*) advised readers to consider buying a rebuilt Jacobs or Wincharger unit.[15] Even wind-energy advocates were cautious. Barry N. Haack advised in a scholarly publication that "on an individual basis, there is no economical advantage to persons installing a small wind electric system under current conditions if there is access to an existing utility network."[16]

Frustrated manufacturers of small machines criticized the federal research effort. Before a congressional committee, Ronald Peterson, board chairman for Grumman Energy Systems, characterized the DOE leadership as "largely sporadic, causing difficulties for us and for Rocky Flats. There has been uncertainty about the budget and operations—on or off, and how much." He said the testing program, which called for placement of two systems in each state, made no sense: some states had virtually no wind resource and others had unusually low-energy costs. Benjamin Wolff, executive director of the American Wind Energy Association, reiterated Peterson's complaints, stating: "Many of our members, particularly those not under contract to the Department of Energy, are frustrated by the apparent unwillingness on the part of the DOE to aggressively promote this cost-effective, renewable source of energy, and its apparent inability to shift from research to application."[17]

DOE engineers had a different agenda. As mentioned in chapter 7, the major effort focused on building a large wind generator—one that would make an impact on the nation's energy mix. To them, small wind-energy converters (SWEC) represented an interesting but unimportant aspect of the technology. For inspiration they looked to Palmer Putnam, not Marcellus Jacobs.

Assessment of the DOE/NASA programs depends on what criteria are used. If we assume that engineers aimed to find out what would not work, it achieved some success. If the goal was to produce a reliable, efficient, large wind turbine, the program failed. In 1983, one expert reported: "In our recent government testing of what is termed first-generation, commercially available machines, 29 of 32 units suffered a major mechanical failure in the first six months of operation."[18] It did not help the reputation of wind energy when the experimental unit at Sandusky, Ohio (MOD O—see chapter 7) failed in less than two days, prompting the *New York Times* to headline its story, "$1 Million for only 30 Hours of Work."[19] The Boeing MOD 2, which sat on a hill overlooking the Columbia River in south-central Washington, did not fare much better. In June 1981, a reporter vividly described how the generator "wrenched its innards." A bit of dirt had jammed the hydraulic system that controlled the blade-feathering device. The blades accelerated to 29 rpm—60 percent over normal speed—and the electrical generator tore apart. The price tag for repairs was one-half million dollars.[20]

MEDICINE BOW

Perhaps the most instructive experiment with large turbines took place at Medicine Bow, Wyoming—a town immortalized in literature. In 1885, novelist Owen Wister visited Wyoming and later featured the cow town in the opening pages of his novel *The Virginian*. The tiny High Plains town has not changed much in one hundred years; nor has the natural environment. Winds still constantly sweep through the town, funneling air eastward from the Great Basin to the Great Plains. In the mid-1970s, representatives of the Bureau of Reclamation and the DOE arrived, looking for wind. By 1979, Lawrence Nelson, project manager for the bureau, spelled out his hope that the agency would eventually put up about fifty wind turbines (representing over 100MW) that would create "enough power to serve 67,300 homes" and save 800,000 barrels of oil a year.[21] The fact that the Bureau of Reclamation (builders of dams and manipulators of water) was involved indicated that a hydropower storage concept, often discussed, would be implemented. This network would be tied into the great power-generating dams of the Colorado River, primarily Flaming Gorge and Glen Canyon dams, with "substantial economic benefits and without adverse environmental effects."[22] During peak wind periods, the electrical energy flowing into the lines could then allow a decrease in hydro production, permitting more water to remain in the reservoirs until needed. Essentially, by decreasing water releases for power production, the wind turbines would create an energy-storage system. These two kinetic energy sources could work well in tandem.[23]

The U.S. Congress agreed to try the plan, funding a demonstration project featuring two large turbines: a Boeing MOD 2 (2.5MW) and a Hamilton Standard WTS-4 (4MW). The two turbines were constructed, raised, and tested, and accepted in 1982. The scale was impressive, even on the vast Wyoming landscape. The tower for the MOD 2 was 200 feet tall; the WTS-4 tower soared to 262 feet. Elevators carried operators to the nacelles. The MOD 2 weighed 580,000 pounds and the WTS-4 topped out at 791,000 pounds. With the blades in a vertical position, the MOD 2 reached to 350 feet above ground and the WTS-4 towered to 391 feet.[24]

For four years, the project generated electricity, but only periodically. Plagued by problems from the start, the Boeing MOD 2 came to a permanent halt after eighteen months, the victim of a burned-out main bearing.[25] When bureau officials received a repair cost estimate of $1.5 million, they quickly terminated the whole experiment. The bureau offered the Boeing for sale, but there were

Boeing MOD 2 (2.5MW) units, like this one at Medicine Bow, Wyoming, were plagued by mechanical problems. In 1987, the Bureau of Reclamation, facing a repair bill of $1.5 million, sold the turbine for $13,000 to a scrap dealer and the tower was dynamited and hauled away. (Courtesy Rick Sorenson, Casper *Star-Tribune*, Wyo.)

no takers. In 1987, a scrap-metal company purchased the unit for $13,000, dynamited the tower, and hauled it away. The WTS-4, at the time the largest wind generator ever built, operated for four years. But in August 1986, a bolt evidently came loose and was ground up inside the generator. The Hamilton Standard was saved from the wrecking ball by the local population, who thought it might function as a tourist attraction. In 1989, Bill Young, an electrical engineer and fervent believer in wind energy, purchased the $10 million WTS-4 from the bureau for $20,000. Young formed Medicine Bow Energy, Inc., repaired the huge machine, and occasionally ran it.[26] It was running on January 14, 1994, when Young heard a collision that sounded like a "clap of thunder." Racing to the window, he saw a seventy-foot section of the blade spinning in the air one hundred feet above the turbine. The accident occurred when the pitch control failed, throwing off the alignment and causing the blade to strike the tower. The next day, 60mph winds caused the turbine violently to rotate back and forth. The four-ton aluminum platform built around the nacelle fell off and by the end of the day the ground at the base of the tower was littered with broken machinery. Needless to say, the WTS-4 was finished.[27]

The Medicine Bow experiment underscored a major drawback to the general concept of economy of scale. When a large turbine broke down, often it could not be repaired. If repair was possible, the cost of bringing in a crane—at approximate rental rate of $10,000 per day—to lift the nacelle off a two hundred-foot tower was prohibitive. Once the nacelle was earthbound and the repair crew could begin work, the idle crane continued to run up expense. Such high maintenance costs might have been mitigated on a large wind farm, with, say, forty 2.5MW turbines. Such an array might have warranted the purchase of a crane. But even with a crane on site, the general operating and maintenance costs would be excessive.

This was the conclusion reached by Pacific Gas and Electric Company executives after they erected a MOD 2 in Solano County near San Francisco. The utility company purchased and operated the Boeing MOD 2 with the idea of testing its reliability. It would indicate the viability—or otherwise—of large wind turbines. If the project proved positive, the company would purchase more MOD 2 turbines, possibly creating a 100MW farm (forty turbines). However, major problems developed, including cracks in the low-speed shaft and difficulties in lubricating the bearing supporting the pitchable blade tips. PG&E lost interest in the project.[28]

In retrospect, the federal research and development (R&D) program of the 1970s and early 1980s did little to promote wind energy. Some wind-energy proponents argued that Congress had cut funding at the time it was most needed.[29] But responsibility for such dire results cannot be assigned to politicians: engineers simply experienced the difficulties of building reliable large turbines. It can be claimed that the federal program enlightened many engineers about the difficulty of successfully building wind turbines. The failure of the first-generation MOD series was not unexpected. They were experimental machines, and only federal research facilities could afford such a large risk. Responding to questions regarding the failure of the MOD 2 and the WTS-4 at Medicine Bow, engineer Bill Young maintained: "When you push the limits of technology and engineering that far that fast, you are bound to have problems." Both Young and David Spera of the NASA Lewis Laboratory have maintained that although most of the machines did not perform as anticipated, much knowledge was gained. Perhaps it can be said that through such experiments engineers learned what not to do.[30]

The federal R&D program should not, however, escape criticism. The early failures can be excused. But were the NASA engineers resilient enough? Were they willing to go in alternative directions? The evidence suggests a certain rigidity. The federal engineers continued to build two-bladed machines, committed to a design that many questioned.[31] Furthermore, they made optimistic predictions without substantive supporting evidence. Federal officials expected the MOD 2 eventually to go into commercial production, in spite of a dismal record of reliability. The DOE predicted that the MOD 5, (the larger 3.2MW turbine) would be in full production by 1986. Twenty would be produced in that year, 50 in 1987, and 112 in 1988.[32] Only one MOD 5 was actually built, and although it has functioned reasonably well (in Hawaii) no manufacturer or utility company has hinted at interest in patents or plans for one. One wind-energy reporter concluded from the experiments of the Wisconsin Power and Light Company that "the assumption that large windmills deliver energy cheaper than intermediate and small windmills has, in our opinion, not been demonstrated. Intermediate and small wind energy systems are more suitable for true mass production."[33] A similar conclusion can be drawn from evidence supplied by the DOE/NASA large systems.

THE DANISH TURBINES

Matthias Heymann, a young German scholar, has examined the American R&D program and compared it with those of West Germany and Denmark. His statistics indicate that, between 1973 and 1988, the United States spent $380.4 million on wind technology, West Germany budgeted $78.7 million, and Denmark was a distant third at $14.6 million. The success record reveals a reverse positioning—a modern parable of the turtle and the hare. The hares, the United States and Germany, put government scientists, large aerospace companies, and university research centers to work on the problem. The results for Germany were, in Heymann's words, "a complete failure" and U.S. scientists, as we have noted, did not fare much better. But diminutive Denmark followed a different path. Capitalizing on its wind-energy tradition (Professor LeCour, and more recently Johannes Juul) the Danes developed a craftsman approach, as distinct from that of engineers and scientists. They had built and installed one hundred reasonably small (50kW to 70kW) turbines in the countryside during World War II, which often provided rural residences with their only source of energy. With the oil embargo of 1973, the Danes fell back on their basic knowledge of wind generation.[34]

The Danish government sponsored some large (750kW to 2MW) projects, but the real success came from idealistic amateurs and craftsmen. These inventors often were working for farm-implement companies. Christian Riisager, a carpenter, sold fifty of his 22kW wind turbines by 1980. A 1978 Danish law required utility companies to accept electricity produced by wind generators (the legislation was similar to PURPA). This inspired the Danes. Investment tax credits further aroused interest, as did a law exempting wind-turbine manufacturers from energy taxes. Wind energy became profitable.

In the United States, the state of California was providing similar incentives and the Danes moved in. By the mid-1980s, the Danish machines dominated the California market. The top wind turbine producer in 1987 was an American company (U.S. Windpower, Inc.), but behind it were five Danish companies (Nordtank, Micon, Bonus, Wincon, and Vestas). In 1987, 90 percent of new installations in California wind farms were Danish-built.[35] A fashion emerged from this competition, often referred to as the Danish design. Generally it featured three blades, upwind design (i.e., the front of the blades facing upwind), medium size, a fixed-pitch hub, stall control, and slower rotor velocity. Above all, it mandated simplicity and very heavy construction. The Danish design captured the market because of weight and toughness, which translated into the magic word *reliability*.[36]

The "slow and steady" approach of the Danes triumphed over the high-tech U.S. approach. Perhaps the American government effort lagged because it was too ambitious. American engineers believed the answers could be uncovered in the crucible of science

Strong and reliable, Danish wind turbines captured much of the California market. Members of the American Wind Energy Association visited the Danish "forest" in 1987. (Author photo)

and high technology. However, as one critic of the U.S. effort noted, building rockets and missiles was much different than designing a wind generator. A missile endures for an hour at most; "windmills have to last—hour after hour—day after day—year after year—producing electricity to compete with the utility company."[37] For comparison, it might be noted that a typical automobile will run 300 to 500 hours a year; a wind turbine is designed to run 7,500 hours in the same period.[38]

The American effort resulted in new knowledge, but it was primarily negative information. It may not have been the "utter farce" that some within the industry believed, but after DOE spent all that money, the legacy was little of tangible value. The one U.S. company that has prospered used virtually none of the technology or features favored by the Lewis Research Center and the SERI (presently redesignated as the National Renewable Energy Laboratory—NREL).

One must not be too judgmental. Wind energy is still in its infancy. Many engineers and wind-energy consultants believe that the Danish turbines will not survive: they are too heavy and inefficient. Robert Lynette, a veteran wind-energy engineer and consultant, views the Danish turbines as a passing fancy. "The technology used by the foreign manufacturers will not be sufficient to survive in the 1990s," he argues.[39] Lynette has faith in the MOD design. In 1994, he was introducing the WC-86, a two-bladed, downwind, teetering hub, lightweight, high-tech, 275kW turbine. The jury is still out.

CHAPTER 9

DECISIONS FOR THE WIND

Amidst all the static forms of the urban landscape,
a colorful wind generator spins with the pulsating
energy of the winds. . . .
[The windmill] has generated a real sense of community pride.

—ENERGY TASK FORCE, NEW YORK CITY, MAY 1977

FOR WIND ENERGY TO PROSPER, THE INDUSTRY NEEDED NOT ONLY RELIABLE machines; reliable markets were needed, too. Neither came easily. Someone outside the industry might assume that finding consumers and establishing a market for wind-power electricity would be a simple task. However, such was not the case. Wind-power manufacturers from the 1920s through the 1950s had found customers for kilowatts in isolated areas, but that market had largely disappeared with the coming of rural electric co-ops and new technology. Furthermore, the new, large, turbines were not directed toward individual users: wind-energy companies would have to feed the utility grid. To do so, some wind-energy advocates and other cogeneration producers convinced, cajoled, and coerced some of the huge utility companies to rethink their ideas of power production. From 1975 to 1985, the nation's huge utility monopolies were compelled by law to accept the power produced by independent operators. Some of these operators used renewable energy sources.

This change did not happen without contention. Utility companies would not relinquish willingly a state-sanctioned monopoly. In the early 1970s, utility companies were entrenched; they were a dominant element of the national economy. Public utility commissioners might manipulate a state's electricity rates, but few, if any, questioned the private-company experts, statisticians and engineers, who controlled and predicted the nation's energy future. Energy forecasters lived in a world of graphs, charts, and models that were incomprehensible to all but professionals. Growth was predictable, at about 7 percent per year, and the utility companies were prepared to meet this increased demand through new, cleaner, and more efficient plants—run on coal, oil, or nuclear. Commissioners might prune profits, but they rarely questioned a utility company's right to monopolize energy production, the method of production, or company predictions.

The beginnings of change happened in New York City and it was an authentic David and Goliath encounter. As the bell for round one sounded, the powerful Consolidated Edison flexed its muscle, denying a group of tenants on the Lower East Side the right to feed windmill-generated electricity into the grid. The saga had begun when a group of environmentalists who called themselves the Energy Task Force and a group of tenants at 519 East 11th Street had won a $177,000 federal grant under the "Adopt-a-Building" program.[1] Tenants contributed plenty of so-called sweat equity into restoring the building, and in their enthusiasm for innovation they decided to incorporate wind energy. Why not install a wind turbine on the roof of their building?. They purchased an old Jacobs 2kW turbine for $1,350 and restored it. By the time it was up and spinning they figured they had made a $4,143 investment. If the Jacobs ran for twenty years, it would produce $11,266 worth of electricity; a thirty-year life would yield $32,356.[2]

But all these figures were based on the assumption that they could hook up to Con Ed power lines. This, was a bold—some might say foolhardy—assumption because it was illegal for independent, customer-owned generating equipment to be interconnected with utility lines. However, the group decided to fight for the right of urban residents to produce and consume their own electricity and to feed surplus into utility lines. To that end, they issued a press release (November 12, 1976) about this new addition to the New York skyline. Such a novelty as a windmill in the Big Apple was bound to draw media attention. It did.[3]

At first, as noted above, Con Ed officials denied the residents access to their lines, but clearly the utility executives were no match for residents. The group pointed out that 98 percent of the energy produced would be used in the building. The remaining 2 percent would be donated to Con Ed—for free. However, they conceded needing to have Con Ed backup electricity. These arguments won the sympathy of the press, local politicians, and the New York State Public Utility Commission. Nothing could please New Yorkers more than bashing Con Ed. Rates had increased 300 percent in the previous decade, and all advertising efforts by the utility giant to win the hearts of its customers were to no avail.[4]

The company was vulnerable and company executives realized it. The following January 18, Con Ed released a proposal recognizing the right of people to produce their own electricity from wind turbines and to tie into the company lines. This was a victory, but it seemed it would be a pointless advance when Con Ed imposed excessive tenant liability on the venture and an absurdly high minimum charge for electricity. At that point, the residents of 519 East 11th Street appealed to the State of New York Public Utility Commission. On May 6, the *New York Times* ran a front-page story headlined "State Tells Con Ed to Buy 2 Kilowatts—From a Windmill." A photograph, accompanying the article, showed residents celebrating on the rooftop and climbing the tower, hands and fists raised in victory.[5] The decision, as an Energy Task Force member observed, represented "the culminating efforts of excellent legal aid, focused neighborhood politics, media interest, a bewildered Con Ed, and a receptive Public Service Commission."[6] Two days later, another New Yorker noted, in a letter to the editor, the broader significance, underscoring that "the decision made it clear that a person may not only generate his own electricity but also sell excess power to his local utility."[7]

The victory had significant social consequences. The windmill project was a community effort of poor, beaten-down people, many of them often unemployed, who became energy producers rather than consumers. They also overcame bureaucratic and legalistic obstacles that were disheartening to all but the most stalwart. Individualism, self-reliance, and a sense of community were all enhanced by the little Jacobs machine, up there spinning on the rooftop. Project leaders put out a booklet that May, declaring:

> Previously unemployed community residents have joined the self-help housing movement to rebuild their neighborhoods. The crippling fear of bureaucracy which so often frustrates the poor is absent on East 11th Street. People there have moved from squatter's instincts to organized, long-range planning for economic growth through sweat equity construction and energy conservation programs. The windmill at 519 represents just a small part of the success they've made at self-help.[8]

These urban apartment dwellers had accomplished a sense of independence and self-reliance that rural people often talked about, but seldom achieved. Robert Nazario, "Mayor of Loisaida (Lower East Side)" caught the significance of it in blank verse:

> The windmill
> wind machine
> to Loisaida's kids
> the "helicopter"
> to East Eleventh Street's elders,
> the "fan"

has many meanings:
an alternative
another way
creation of electricity for one's household
removal of dependency on Con Ed
knowledge that Con Ed cannot simply wade in and lock up your windmill
Now, small people, poor people, can proudly say
There are three legal, recognized power companies in New York City: Con Edison, Brooklyn Union Gas, and the people of 519 East 11th Street.[9]

From the point of view of energy supply, the amount of power produced was insignificant. But the wind turbine was a symbol: people need not be entirely reliant upon centralized utility monopolies. In Congress, Senator Edward Kennedy noted the struggle, calling the Jacobs "the little windmill that could."[10]

About two months after the tenants' victory, Con Ed suffered a devastating blackout. Much of the city was without power for as long as twenty-five hours (July 13, 1977). Unlike the carnival atmosphere that is associated with the Northeast blackout of 1965, this power failure prompted rioters and looters to go on a $50 million dollar spree.[11] This vulnerability to the centralization of power led many people, already aware of the diminutive windmill, to think there might be another way: one that would not only be economic but in harmony with the environment (including the environment of a vast city).

To most of the powerful utility companies, the few independent producers of energy were annoyances, not to be taken seriously. In 1977, Representative Henry Reuss, of Wisconsin, who had installed his own wind generator near his home at Chenaqua, Wisconsin, spoke for many wind-power advocates when he remarked that many utility companies "are displaying a public-be-damned attitude at its worst."[12] Most utility companies did not deny the right of the individual power producer, for it would be politically unwise to oppose a renewable and nonpolluting power source. However, their policies were obstructionist. One resident near Casper, Wyoming, had to threaten to take his case to the local television cameras before the power company would work out a reasonable contract. A Wisconsin woman, who vented her frustrations in a letter to her local newspaper, protested that "since our [wind] generator has been up we have received nothing but grief from our power company. This has ranged from threatening phone calls and letters, harassment and out and out muscling. They offer contracts for us to sign that are totally one-sided. They want to literally steal our excess power and threaten to turn our power off if we don't cooperate." She and her husband had "invested several thousand dollars" of their own money. The Wisconsin Electric Power Company denied the charge, claiming that they could not allow the meter to run backwards since the rates on power sold and power received were different.[13] Similar power games were played out in many places across the country.

Some companies were obliging. A spokesman from Ohio Edison explained that his company "will operate in parallel with customers provided they have the right equipment" and similar positions were taken by power firms in Alabama, Connecticut, Iowa, Maine, Michigan, Minnesota, Rhode Island, and Virginia.[14] These were the more liberal companies and states. Many simply distanced themselves from the tiny producers, refusing to buy the excess power generated.

Much of the bickering revolved around the concept of "avoided costs," but in general utility companies became defensive because they were loosing their traditional monopoly on production. The annoying gnats that Samuel Insull swatted in the 1920s were becoming more numerous and more insistent. The entrenched monopoly arrangement, hammered out during the presidency of Theodore Roosevelt, began to crack.

CALIFORNIA GREEN

The crack became a chasm in California, a place where wind energy would prosper on a grand scale. The state had always been mythic—a place of legends, ever on the cutting edge of progress. It had led the nation in fashion trends, freeways, lifestyle choices, population, and a whole range of social and economic criteria. Historically, the state derived its name from the fertile imagination of a Spanish novelist who, in about 1510, wrote of a mythical island, inhabited by Amazonian women and blessed with abundant gold. This place was called California. The Amazons never materialized, but in 1849 the gold did. Fifty years later, "black gold," oil, would add to the prosperity of the state. Now, in the 1990s, an unlikely resource, the wind, is playing an increasingly important role. On the mountain passes, the deserts, and the hills of California, thousands of turbines generate some 90 percent of the wind energy produced in the United States.

Why has the state taken such a commanding role? California has advantageous winds, but no more so than other states. Perhaps it was the environmental climate of the early 1970s. California was a land of paradoxes: many residents praised the state's top tourist attraction, the sun; other, in increasing numbers, damned her most infamous feature, the smog. The U.S. Bureau of Census had acknowledged California as the most populous state in the Union on July 1, 1964, but the growth was a two-edged sword.[15] In 1966 Richard Lillard pubished his hard-hitting fact-packed book about Los Angeles and pollution, *Eden In Jeopardy*; and in 1965 ecologist Raymond Dasmann sounded an alarm about the whole state in his stirring *The Destruction of California*.

Politics reflected this environmental concern. The years with Ronald Reagan as governor had been prosperous, but at a price. Many Californians who voted in Jerry Brown (1975–1983) believed California had gone too far. Was *biggest* necessarily *best*? Did growth equal progress? And what was progress? Perhaps, they mused, human health was more important than a healthy economy. Urban dwellers were the most concerned. They questioned the substitution of the artificial (Disneyland) for the natural (open space). They pointed to the mediocrity of the suburbs and the vulgarity of neon-lighted cities. Californians' propensity to leave their cities on weekends led historian-sociologist Lewis Mumford to ask: "If the places we live and work were really fit for human habitation, why should we spend so much of our time getting away from them."[16]

The state's environmental problems mounted, but fortuitous leadership emerged from environmental groups. Organizations such as California Tomorrow and the influential Sierra Club sounded a note of hope rather than painting scenarios of horror. One of the hopeful alternatives involved the state's energy future. Would the burgeoning state choose a future of more and more huge generating stations—nuclear, oil, and coal-fired? Or were there other avenues? The election of Jerry Brown opened the debate. By 1980, a revitalized California Energy Commission offered two scenarios, clearly favoring the one that stressed "inner-directed" concerns: ecological balance and integrity, attitudes favoring cooperation over competition, and a "reduced emphasis on material gain and the material aspects of status."[17]

In the early 1970s, this second set of values was not evident in the board rooms of the two great California utility companies, Pacific Gas and Electric (PG&E) and Southern California Edison (SCE). In regard to the *production* of energy, they took no innovative or meaningful steps toward change. On the consumption side, PG&E and SCE encouraged individual conservation actions. Tips on how to save energy became commonplace, and advertising efforts switched from an ideology of abundance and consumption to one of scarcity and conservation. The companies sponsored energy audits of consumer homes and distributed information on home insulation and passive solar systems.

Thus, the utility companies appeared to be doing their share of improving the environment by energy belt-tightening. But all their efforts focused on consumer conservation and executives still assumed that demand would increase. Historian Richard Hirsh described a smug, self-confident mood in characterizing PG&E, the huge northern California utility company: "An industry giant, with about 10,000MW of capacity in 1970, PG&E entered the decade with old strategies intact. Expecting load growth to continue at its traditional rate of 6 to 7% annually, the company foresaw the need to construct five new nuclear power plants for use in the 1980s at a cost of about $13 billion."[18]

Neither PG&E nor SCE executives questioned such growth projections. In light of California's staggering population increases, their effort went into considering how to meet future demand. They found encouragement in a 1973 study by the state of California, *Energy Dilemma*, that concluded that "if the demand for electricity is to be met, a number of new nuclear plants need to be built."[19]

In 1975, a more comprehensive study was prepared for the California State Assembly by the Rand Corporation. The assembly asked Rand (a creative think tank) to "identify the major energy issues affecting California, to assemble relevant factual information bearing on these issues, to define the key alternatives that the state could pursue, and to discuss the implications of these alternatives for state energy policy."[20] With such a broad mandate, and with free rein to suggest innovative solutions, the Rand thinkers might have come up with ingenious and imaginative ideas, putting to shame the uninspired intellects of PG&E, SCE, and state officials. To the contrary, the Rand report stated that neither wind nor solar generation "will contribute significant amounts of electricity before the end of the century" and "nuclear generation facilities [will] play a major role in future plans for electricity generation capacity for the state."[21] The report dismissed wind energy as a technology so fraught with problems that meaningful use of that resource would be "beyond the time horizon of this study [i.e., beyond the year 2000]." In the report's conclusion, ten additional sources of energy for California were discussed.[22] Wind power was not among them.

With the advantage of hindsight, we now can see that the Rand people did not have their creative environmental hats on. Perhaps the early wind-energy failures convinced them that engineers would never resolve the technological problems. Perhaps the legislature simply wanted a report to validate a majority view, and the Rand people gave it to them. That the Rand team ignored wind-energy potential is understandable. But it is difficult to fathom why the report did not anticipate the most important new energy source of the 1980s: *cogeneration*.

Even as the Rand think-tank team wallowed in the security of precedent, two pages of rather revolutionary material were being slipped into a 1975 California Public Utility Commission (CPUC) report. The basic business of the report was routinely to grant to PG&E a rate increase the company had requested. According to the report, the commissioners regarded conservation as "the most important task facing utilities today." Furthermore, the "unchecked proliferation of power plants" could not be allowed to continue. A new CPUC, buttressed by two new appointments (Governor Brown had just taken office) called for conservation and alternative sources of energy.[23]

Three environmentalists, working in a vacated Berkeley fraternity house, took note of this rather radical CPUC pronouncement. These professionals, staffers for the Environmental Defense Fund (EDF), would challenge the utilities' notion that new generating plants, using nuclear, oil, or coal, must be built to meet increasing demand. Zach Willey, a computer wizard, and Tom Graff and David

Roe, both lawyers, argued for a new approach before the California Public Utility Commission. They believed—and Willey's computer model provided the data—that PG&E could save the environment, and a great deal of money to boot, by not building the projected nuclear plants. Increased demand would be met by conservation, not increased capacity. *There was a way to do it.* The California utilities need not advance in traditional lockstep, increasing capacity. They could *decrease* demand.[24]

Could conservation really be attractive to PG&E? As attractive as building new plants? The answer for Willey, crunching numbers in EDF's Berkeley garret, was to allow the company to profit from conservation. PG&E could finance insulation, water heaters powered by solar, and efficient lightbulbs, all at a financial loss; however, the loss would be recouped by adding the investment in conservation to the CPUC rate structure. In adding its investment to the rate structure, the utility might allow for a reasonable profit, around 12 percent. Thus, a utility company might profit by *not* producing electricity. "To be fair, [conservation] ought to be treated exactly like an investment in [a] power plant," Willey argued. "If PG&E makes the investment, they should be able to put it into rate base and earn a profit on it, just like any other investment they make."[25] From a capitalist, stockholder point of view, it did not make any difference if profits came from decreased consumption or increased production. But the California environment would be the sure winner if, as EDF argued, no new power plants were required.

The appetite for growth and the habit of continued building programs did not fall away easily. A phalanx of PG&E lawyers fought the minuscule resources of the EDF, challenging Willey's models and accusing EDF of "a concerted effort to distort, confuse, and twist the record."[26] Flexibility was not the forte of the nation's largest energy supplier. Only a hefty fine from California's PUC and ongoing embarrassment of their lawyers and computer experts stirred the lethargic company leadership into action.

Assistance came on the political front. Leo McCarthy, the powerful speaker of the California Assembly, embraced the conservation idea—not construction. By 1977, Brown's administration declared in favor of utility company investment in conservation and renewables such as solar and wind.[27] Meanwhile, David Roe and other EDF lawyers continued to hammer away at the PG&E and SCE contention that new power plants would be required for the 1980s. The greatest battle involved the Allen-Warner Valley power plant, a $5 billion joint venture of the two companies. Pummeled by the EDF lawyers, economists, and Governor's Brown's administration, PG&E and SCE finally withdrew their application to build the plant. Although they proposed to resubmit the application, it did not resurface. As Roe wrote: "PG&E and Edison went back to their drawing boards and figured out how to meet all their electricity needs through the end of the 1980s without ordering a single new monolith."[28]

By 1980, the soft energy path—a path of profit through conservation—had conquered California. Both PG&E and SCE accepted the idea that new demand could be met by cogeneration and renewables. Wind energy was the most prominent of these sources. For the utilities, there were tremendous savings of capital expenditure. The downside for the utilities was the tacit admission that the companies no longer monopolized the production of electrical energy: to some degree, they became purchasers and distributors of energy. In 1981, a writer for *National Geographic* found that PG&E people viewed themselves "more as energy-service specialists rather than simply power providers" and SDE acknowledged renewable energy to be a "preferred technology."[29]

Some of this remarkable turnabout can be credited to the EDF effort, but the kudos must be shared with dedicated Brown staffers in Sacramento. From 1975 through 1983, California was headed by young Jerry Brown—Governor Moonbeam, as he was sometimes called. Brown favored an energy policy that earned the label of "woodchips and windmills."[30] Whatever the appellations with which he was saddled, during his eight years of office Brown moved the state in new and novel directions. Minorities, women, global issues, and renewable energy took center stage. There is little evidence that Brown was enthralled by wind energy, but he was persuaded that renewables could play a role in the future of the state.

George Stricker, former California Wind Energy Association president, recalled that the governor surrounded himself with "whiz kids, Ph.D. progressive types" who wanted to make a difference. "Sacramento just became a hot-bed of renewable energy ideas."[31] To explore the possibilities, Brown revitalized the California Energy Commission (CEC), which had been moribund under Reagan. By 1980, CEC employees were churning out reports and contracts at what some thought a frantic pace. Robert Thomas, wind-energy program manager for the CEC from 1980 through 1982, recalled that the "period was one of those unique moments in time, which everyone in our group secretly understood. We were drawn into a highly charged and dynamic situation that we all responded to beyond our normal capacities. We knew somehow that we were participating in History."[32]

Part of this dynamic situation involved a new agency: Brown set up an Office of Appropriate Technology, appointing whiz kid Sim Van der Ryn as head (see chapter 7). Around him, Van der Ryn gathered a staff committed to alternative energy solutions: there were specialists and consultants on community gardens and composting, on solar strategies, on sewage solutions, and—as noted earlier—on wind. This team advised people working on projects around the state. Ty Cashman, as we have seen, set up the "wind desk," providing technical and financial advice for anyone in the state contemplating building, or erecting, a wind-energy unit.

As it turned out, Cashman's major contribution came in the political field. The state legislature was working to implement the new mood of conservation that was abroad in the state. Investment incentives through state income tax credits were being considered. Such credits for new projects would be particularly attractive to the wealthy, who might, for example, install a $5,000 solar heater for a swimming pool and write off 50 percent ($2,500) of the investment.[33] Under this legislation, introduced in 1978 by Senator Gary K. Hart,[34] tax credits would be given to solar installations. Wind energy was not mentioned. When Cashman noticed the omission and called it to the attention of David Modisette, Hart's legislative assistant, Hart agreed to amend the bill. It thus included a phrase saying that tax credits would be given for "wind energy equipment for the production of electricity or mechanical work."[35] After a battle in the senate ways and means committee, in which chairman Willie Brown opposed a credit for wind energy, it barely passed. Once signed by the governor, this energy tax-credit law allowed a 50 percent credit for small systems and a 25 percent credit for industrial systems. The 25 percent credit, combined with generous federal tax credits, created the financial climate that kindled the remarkable growth of the wind-energy business in the early 1980s.[36]

Federal and state tax credits provided an incentive for investors to build wind farms, but who would buy the electricity? The federal government resolved that problem with the passage of the Public Utility Regulatory Policy Act of 1978, referred to as PURPA. This act reflected the effort of the Jimmy Carter administration (1977–1981) to move the nation toward energy independence. Within one hundred days of his inauguration, Carter had sounded the alarm, declaring the nation was running out of gas and oil—a situation that he said was both a crisis and a mandate, so serious as to be "the moral equivalent of war." Much of Carter's appeal was given a contentious reception and Congress soon tied up his energy package. However, amid the threat of another OPEC embargo and astronomical prices, possibilities for renewable energy made headway. Public awareness reached a peak on May 3, 1978, when fifteen million Americans participated in Sun Day, a national event celebrating a coming Solar Age.[37]

THE PURPA REVOLUTION

The Commission shall prescribe . . .
rules . . . to encourage cogeneration.

—PURPA ACT, 1978

It was a much less heralded event that made possible the tremendous advance in California wind-energy production: PURPA was making a tortuous way through Congress as one of the five parts of the National Energy Act of 1978. For the average person, PURPA represented complex formulas on rate structures that bordered on the incomprehensible. It was a piece of legislation that reflected utility company interests, and would have been unmemorable except for section 210, titled "Cogeneration and Small Power Production."[38] Section 210, although it represented "the most underlobbied, unsung piece of the National Energy Act," would change the nature of the nation's energy production.[39] Environmental historians have long considered the uncontested Forest Reserve Act of 1891 as one of the most significant acts in U.S. environmental history, but in time the unopposed passage of section 210 will be considered a watershed event of equal stature.

The section required that the Federal Energy Regulatory Commission prescribe rules for "cogeneration" and "small energy power production." *Cogeneration* referred to relatively small turbine generators (gas or steam) that could utilize discarded heat used for processes such as papermaking or petroleum refining. *Small energy power production* largely referred to solar sources of power (solar, wind) or hydropower. The guidelines required utility companies to sell electric energy to qualifying cogeneration facilities and, most significantly, *to purchase electric energy from these facilities*.[40]

Utility companies might have evaded such a commitment by establishing a ridiculously high rate structure, but section 210 required that the rates "shall not discriminate against qualifying cogenerators or qualifying small power producers." According to the legislative analysis, the drafters included this section because of concern "that the electric utility's obligations to purchase and sell under this provision might be circumvented by the charging of unjust and non-cost based rates for power solely to discourage cogeneration or small power production."[41] What should be the rate? Section 210 stated that the utility company must pay to the cogenerator "the cost to the electric utility of the electric energy which, but for the purchase from such cogenerator or small power producer, such utility would generate or purchase from another source."[42] This "avoided cost" included the "fixed and running costs" that a utility company can avoid by a purchase program. These expenses would be broken into two types: (1) energy costs (fuel, operation, and maintenance); and (2) capacity costs (the capital costs involved in meeting "peak demand" periods).[43] Of course, the establishment of the avoided cost rate is a matter on which opinions diverge—an area of compromise and a search for equity between company and public.

The PURPA legislation initiated a small revolution in U.S. power production. First, the legislation invaded the power-production monopoly of the utility companies. Second, it signalled that the utilities should think about smaller means of production. Third, the act encouraged the use of renewable forms of energy production. In fact, the law stipulated that "qualifying facilities" (QFs) must have as their primary power source biomass, waste, or renewable resources (including water, solar, and wind).[44] Fourth, the act avoided stifling bureaucracy by exempting small producers from the jurisdiction of the Federal Power Act. Fifth, the act did not define rates, but the intent was clear: cogenerators and small producers of electricity must be treated fairly by the utility companies and the state public utility commissions with which they would have to deal. Sixth, PURPA gave notice that, with regard to electricity generation, the traditional regulated monopoly model made little sense, since, as one writer put it, "no vestige remains of the economies of scale that traditionally were invoked to justify awards of exclusive generating franchises."[45] Finally, the act clearly placed the resources of the federal government in the camp of renewable energy. Before PURPA, the government restricted its activities to research and development; now it would take a direct leadership role in changing the course of the nation's energy future. For wind energy, PURPA swung wide the gates of opportunity.

Although the idea of independent power slipped through Congress, utility companies soon mounted concerted attempts to subvert its intent. Particularly when the price of oil dropped—the mid-1980s—the utility companies wished to renegotiate generous "avoided cost" contracts with "floating-rate contracts," claiming their responsibility to protect their customers (ratepayers).[46] In 1986, at least one utility called for a total repeal of the act. However, pleading "the welfare of customers" did not ring true, particularly since utility companies across the land were attempting—some with success—to have customers pay for failed nuclear projects that would never produce a kilowatt-hour of electricity. The utility company opposition to PURPA was driven by forces other then customer protection. For the first time in their history, the utility companies lost their total control of power production. With the law nudging or shoving them forward, they had to negotiate agreements for the purchase of power. In this respect, PURPA opened doors. PURPA diluted centralization. Entrepreneurial possibilities opened up, challenging monopoly with a concept of shared responsibility.

Jan Hamrin, executive director of the Independent Energy Producers Association, stated it well when she told a Senate committee in 1986 that "the issue here is that utilities do not trust anything they do not own themselves."[47] Hamrin succintly went on: "PURPA is that rare piece of legislation which through its creativity and flexibility is able to achieve the results envisioned by its sponsors." The environmental logic of PURPA, combined with its use of market forces and competition, allowed it to withstand attack and provide the wind-energy industry with an authoritative legal instrument.

CHAPTER 10

CALIFORNIA TAKES THE LEAD

Wind-energy people were "building something together—
like the Amish at a barn raising.
It was a heady experience.

—PAUL GIPE ON THE YEARS, 1983 TO 1985

WHEN HISTORIANS EVALUATE THE ENERGY HISTORY OF THIS ERA, THE unheralded PURPA legislation will loom large. Not only did it open up the energy business to small entrepreneurs, it stimulated new forms of energy production, particularly those using renewable sources. Monopolies could be challenged successfully by small groups, such as the residents of 519 East 11th Street. Most of the new energy producers were groups much larger than the New York windmillers. These interests were lured into the game not only by the new market but also by the prospect of making a profit, provided they could produce energy at less than the "avoided cost." Of course, there was no *promise* of profits, but a combination of the PURPA Act, state incentives, federal tax credits, plus accelerated depreciation on the turbine units almost assured affluent investors against loss. It was an attractive proposal for investors—and thus took shape in California the most ambitious wind farm projects the world had seen.

The State of California created a hospitable atmosphere, but what happened in California must be set in a national context. In the 1970s, OPEC and foreign oil emerged as a threat to U.S. independence. The decade had provided a series of embarrassing if not humiliating international incidents, many of them associated with oil. As a result, wind energy or any other form of energy production might provide a Band-Aid for a wounded national pride. In arguing for the Wind Energy Systems Act of 1980, one congressman proclaimed that voting for the act would "send OPEC a message that we are not going to continue to depend on their narcotic." Development of wind energy would help to end dependence on "OPEC's needle."[1]

Rep. Richard Ottinger (New York) and some other congressmen were more disturbed by diminishing energy sources. Wind-energy development was not only long overdue, it represented one more aspect of the painful passage from abundance to scarcity. Ottinger told the House: "We are moving from an era of very cheap energy to one where all our alternatives for providing energy will be very expensive."[2] Such a transition would call into question not only economic growth but the fundamental precepts of progress. Utility companies (in general, supported by the public) had subscribed to the notion that a "no growth" scenario would threaten the nation's economic welfare. To halt the upward trajectory of the national energy consumption curve would be the very antithesis of progress. Yet, the old "progress" (huge power plants and profligate consumption of fossil fuels) was dead, killed by the Environmental Defense Fund attorneys and Amory Lovins's advocacy of soft energy paths. The public was increasingly critical and sensitive. Being on the receiving end of criticism and energy restructuring was a difficult pill for utility company executives to swallow. Yet swallow it they must. Historian Richard Hirsh advised: "Utility managers . . . must give up their old values about growth—at least for the moment—and learn to do well with extant technology."[3]

This new sensitivity also permeated state regulatory agencies and state legislatures. In California, the Public Utility Commission

(PUC) members and staff supported the development of solar and wind resources and they assured their objective through a generous interpretation of "avoided costs." PURPA guaranteed a market at avoided cost rate, but just what should the rate per kWh be? The question was crucial—and it had to be decided separately by each state public utility commission. For instance, New York established high avoided cost rates, but the legislature did not provide tax credits. The commissions in Washington, Oregon, Idaho, and many state of the Midwest established a rate of $0.01 to $0.02 per kWh. This very low rate practically guaranteed that wind-energy development would not occur. The California PUC, on the other hand, established a rate of approximately $0.07 per kWh.[4]

The California legislature was also supportive. Legislative interest can be traced to the 1978 Mello Act, named after Rep. Henry Mello. The act appropriated $800,000 to the California Energy Commission to accelerate the commercialization of wind technology. The Mello Act set ambitious goals. By 1987, 1 percent of California's electrical energy would be produced by wind generators. By the year 2000, that figure would jump to 10 percent. Perhaps 10 percent sounds modest, but it would represent 7,000MW of power—output equal to seven average-sized nuclear or coal-fired plants. To put it in other terms, if, in the year 2000, the utility companies were still generating power with oil, a 10 percent wind array would save fifty million barrels a year.[5] By this extraordinary effort, the state hoped to lessen air pollution, help control energy costs, and create new jobs. The legislation assumed that many of the systems would be built in California.

Remarkably, state officials expected wind turbines to increase "the reliability of the state's electricity supply." Reliability has rarely been a substantial selling point of wind-energy systems, but the experts advanced the argument that "since wind is not subject to oil embargo, coal strikes, natural gas diversion, pipeline disruption, nuclear moratorium or shortages of water, this investment would help to stabilize the electricity supply picture."[6]

Well funded by the legislature and with an ambitious production target, the California Energy Commission (CEC) swung into vigorous action. In a flurry of activity, the commission produced wind-energy reports—the data essential for development. California Wind Energy Association director George Stricker later recalled that these reports were highly prized. Company representative "would rush for the reports as soon as they came off the press." Mike DeAngelis, another CEC staffer, also praised these studies, asserting they were "one of the main reasons California has almost 90 percent of all wind development worldwide—it could have gone elsewhere in the United States but the data were just not there."[7] Other states did provide enticements for wind-energy development (for instance, by comparison with California's 25 percent tax credit, Arkansas offered 100 percent and Oklahoma and Oregon both offered 35 percent), but investors found generous tax incentives of little value without research on a state's wind potential.[8]

Many of the early California wind surveys produced encouraging figures. The state had more wind than expected. Previous estimates were based on historical data recorded at airports, but as energy expert Christopher Flavin noted, "these sites are often chosen specifically because they lack wind."[9] Once meteorologists moved into the mountain passes, they found a new goldmine. One study of wind potential estimated that California's open-range land offered a total of 527 billion kilowatt-hours per year. To put this figure in perspective, it represented almost 1.3 times the total electrical energy demand forecast for California in the year 2000—or approximately 25 percent of the nation's electric consumption in 1980.[10] James I. Lerner of the CEC, testifying in 1979, was most upbeat. The CEC wind assessment "gives us confidence that the wind that is available for electrical generation is 7 to 10 times larger than what we will need to reach a 10-percent target."[11]

The CEC not only recommended areas of wind potential; it also suggested how the electrical energy might be employed by the state's five major utility companies.[12] Since electrical energy cannot efficiently be stored, it must be generated when needed and used immediately. However, an indirect method of storage is possible: wind energy can be used in conjunction with hydropower. When the wind blows, water normally used for power production is held in storage behind the dam. In some cases, dam supervisors can utilize the electric power for pumps to lift water from below the dam back into the reservoir. When the wind diminishes, the supervisor can increase the water flow, maintaining the preferred average flow from the dam. Thus, hydroelectric facilities can serve as a back-up for when seasonal wind lulls occur.[13] A 1975 study by the Federal Power Commission identified a number of potential "pumped hydro sites," each with 1000MW capacity or more. The report put forth the astonishing figure of 144,000MW of potential pumped hydro within California. At an 80 percent availability rate, this would mean 115,000MW of energy. To convert all of this potential energy would require 29,000 4MW wind generators.[14] Neither the number of wind machines nor their size (4MW) was practical, but it was a good shot at estimating available capacity for a future generation that might face scarcity of natural gas, coal, or oil.

With the resource well defined, the CEC provided convincing data that the wind could be harvested economically. Large wind turbines would generate electricity for three to four cents per kilowatt-hour, somewhat lower than electricity from conventional power plants. Utility companies, both investor owned and publicly owned, should encourage the new technology, particularly since the wind-energy resource peaked in the summer months, when air conditioners demanded electricity.

Perhaps the most encouraging aspect of the CEC presentations was the assertion that wind power systems were "benign environmentally." "They do not emit air pollution nor consume water for cooling," stated a CEC report. "Aesthetics and land use do not appear likely to limit the deployment of this technology."[15] The first assumptions—about pollution and water—proved to be correct: in the semiarid and arid West, which seemed always to be in draught or under threat of drought, the water factor was especially welcome. But the predictions about aesthetics and land use were off the mark. These considerations would take center stage and limit wind energy's potential.

INCENTIVES

To realize these ideas, the cooperation of the major utilities was of course necessary. But it was private companies and private capital that responded most enthusiastically to the state's ambitious goals. Dramatic amounts of capital were needed. The CEC estimated that $300 million would take care of the 1 percent objective. But $5 billion would be necessary to meet the year 2000 objective. Obviously, such a risky technology would have to offer enticing rewards. These incentives came in the form of federal and state incentives—incentives so generous that many journalists and critics soon labeled the wind farms as tax farms.

At the federal level, in 1981 a corporation or investor in a limited partnership could enjoy two tax credits: the Regular Investment Tax Credit (RITC) and the Business Energy Investment Credit (BETC). The RITC provided a 10 percent credit on eligible costs; the BETC offered a 15 percent credit. Utility companies could not qualify for the BETC, but wind-energy companies and individual investors could. Thus, for the wind-energy company or the individual investor, 25 percent of investment could be immediately recouped. If the tax credit could not be applied in the initial year, it could be carried forward to future years.[16]

Between January 1978 and December 31, 1986, California offered lucrative tax credits, equalling or exceeding the federal incentives. Wind systems costing more than $12,000 and not installed at a home were eligible for a tax credit of 25 percent, with no limit on the state contribution. This credit could be passed on from the owner/developer to the purchaser. Significantly, this was in addition to the federal tax credit; hence, there was a total state and federal tax credit of 50 percent. For small wind turbines in residential or farm use, the state provided a 55 percent credit, to a limit of $3,000.[17] Total federal and state tax credits for residential wind systems could not exceed 55 percent.

In addition to tax credits, a number of loans and loan guarantees were available through the Department of Energy, the Farm Home Administration, the Department of Housing and Urban Development, the Farmers Home Administration, and the Small Business Administration (SBA). Most of these loans were for relatively small projects, with the intention of helping individuals finance solar projects. However, the SBA offered energy loans of up to $350,000 to small companies engaged in wind-machine manufacturing, distributing, and servicing. Between January 1979 and May 1980, this agency made fifteen direct loans in California, totaling $3,466,000, and five loan guarantees, totaling $858,000.[18]

The California Alternative Energy Source Financing Authority Act, which took effect on January 1, 1981, provided a further incentive. The act allowed corporations and partnerships engaged in alternative energy projects to issue bonds (up to $10 million for an individual project) that would be exempt from state taxes, the interest rate on such bonds thus being lower than that on taxable bonds.[19]

The tax-credit package, combined with other financial incentives, acted as a catalyst, transforming idea into reality. On the individual level, many Californians were encouraged by the Office of Appropriate Technology, which publicized that fossil fuel costs "make the production of electricity from wind energy one of the most attractive and cost-effective alternative generation technologies currently available." The office admitted that some people found the turbines noisy and unattractive, but such concerns were insignificant when measured against the "sense of community accomplishment which comes with local energy production."[20]

1983: THE BOOM

[Tax credits] helped create the fad of the year: the wind park tax shelter.

—*FORBES MAGAZINE*, JULY 18, 1983

The big rush, however, came on the commercial front. Suddenly, wind-energy development became trendy in investment circles. Wealthy investors, with $50,000 or $100,000 to disburse, called their brokers, captivated by the idea of aiding the environment while padding their pocketbooks. The California Energy Commission, perhaps with hyperbole born of enthusiasm, encouraged small investors, stating that "virtually anyone can invest in wind energy. In California the development of major wind energy projects has been initiated by investor-owned and municipal utilities, as well as by individuals and private developers."[21] A *Forbes* magazine writer caught the gist of the excitement when he wrote that all these tax credits "helped create the fad of the year: the wind park tax shelter."[22] It was a time when "limited partnerships" were popular, and wind energy proved to be one of the most attractive areas for such activity.

Fairly typical was the "Windmaster Partners 1983-1," a $14.2 million proposal made in June 1983 by BEPW Development Corporation, "a wholly owned subsidiary of Blyth Eastman Paine Webber." The corporation offered ("only to residents of California") the opportunity to be a limited partner in a WindMaster Model 200-22 turbine by purchasing one or more units ($100,000 per unit).

The prospectus offered tremendous incentives. A $100,000 investment would realize tax credits of $52,366 the first year. Over the first five years, the total tax write-off in tax credits and depreciation (five-year life) amounted to $89,510—almost 90 percent of the investment. Over the projected fifteen-year life of the project, the potential cash return was $137,765.[23] A month after the initial offer, BEPW sweetened the proposal. Potential investors received a revised summary of benefits: a five-year tax saving of $107,378 and a fifteen-year cash return of $128,697.[24] Investors responded well to this prospectus. By 1992, HMZ-WindMaster (a Belgian turbine manufacturer) had installed 129 of the 200kW machines, 30 250kW, and 15 300kW. However, elimination of tax credits for solar renewables decreased the tax savings and the projected cash return has not materialized.[25]

Few, if any, of the companies paid-off in cash, but it did not seem to matter. In promoting its wind-energy investment, the Oak Creek Energy Systems Inc. stated that "on the surface, it may not matter which wind turbine developer you choose." The "tax savings nearly total the entire cost of your wind turbine." The cash dividends on its turbines would represent pure profit.[26] By 1981, thirty-eight companies, some old, many new, had maneuvered to lease land, raise money, purchase land, and sign agreements to sell power with PG&E or SCE. Although the California Energy Commission had located "potential wind resource regions" in every county of the state, three sites proved most attractive.[27] In northern California, the historic Altamont Pass funneled travelers from the Bay Area to the northern San Joaquin Valley and south to Los Angeles; it also served as a conduit for wind. The vacuum caused by the rising hot air of the Central Valley pulled the cooler coastal air over the hills. Not only did Altamont provide an excellent wind resource (mean annual wind speeds of more than 14mph), the region's undeveloped rolling hills were utilized only by ranchers and their livestock. Most ranchers willingly leased land, seeing wind-energy development and grazing as compatible—not to mention the money. Hugh Walker received about $300 to $400 per year per windmill. With one hundred wind turbines on his land, this was a welcome $35,000 a year. Some ranchers took satisfaction that "all that wind was good for something besides blowing my hat off." Now, stated another rancher, he wouldn't "cuss it so much."[28]

Two southern California passes also possessed excellent winds and available land; both draw inland air to the Mojave Desert. Tehachapi Pass, southeast of Bakersfield, is one of the windiest places in the world. Afternoons, heating from the nearby desert sucked the Central Valley air eastward, sometimes at gale force in the spring and early summer.[29] Further south, San Gorgonio Pass is flanked by 11,000-foot-high San Gorgonio Peak to the north and 10,000-foot Mt. San Jacinto to the south. It is a narrow, natural wind funnel and has been recognized as such since the days of Dew Oliver. Nestled nearby at the base of Mt. San Jacinto is the resort town of Palm Springs. As discussed in chapter 11, the residents have had ambivalent feelings regarding the some four to five thousand wind generators that saturate the pass.

California has many other windy regions, but land availability and land-use decisions have often prevented development. The Boulevard area near San Diego witnessed some development in the early 1980s. This has since failed. Pacheco Pass, the primary thoroughfare from south San Jose into the San Joaquin Valley, has excellent wind but little available land.[30] The potential is there, however, for wind/hydro development, using the San Luis Reservoir. The only other wind farm of size is at Carquinez Straits in the Delta Region of northern California, where U.S. Windpower installed over six hundred wind generators. More are scheduled.

As landowners leased acreage, the business quickly acquired a global character. Germany, Belgium, Denmark, the Netherlands, Ireland, Japan, and the United Kingdom all got in on the California wind boom. Companies with unfamiliar names—HMZ-WindMaster

Members of the American Wind Energy Association visit the U.S. Windpower (now Kenetech) facility at Altamont Pass, California, in 1987. Altamont Pass was the world's first large-scale wind-energy generating station. (Author photo)

(Belgium), Vestas (Denmark), Micon (Denmark), Bonus Energy (Denmark), Nordtank (Denmark), James Howden (United Kingdom), Wind Energy Group (United Kingdom), Stork FDO (Netherlands), and M.A.N. (Germany)—shipped their wind turbines to California and set up shop, hoping to glean knowledge as well as profits from this new bonanza of opportunity.[31]

American companies also prepared to capitalize. Fayette Manufacturing Corporation, a small Pennsylvania-based wind turbine manufacturer established in 1978, moved its operations to Tracy, near Altamont, in 1981, eventually installing some 1,709 turbines in Altamont Pass wind parks.[32] FloWind (a sister corporation of Flow Industries, Inc.) incorporated in March 1982, obtaining rights to manufacture the seventeen-meter vertical-axis wind turbine originally developed by Sandia National Laboratories in Albuquerque, New Mexico. The company operates 148 of these unusual 150kW "eggbeater" machines at Altamont, as well as twenty-one of a larger (250kW) type. These graceful FloWind turbines are also found at Techachapi.

By far the most important American wind energy company at Altamont Pass is Kenetech Windpower (formerly U.S. Windpower). The company has approximately four thousand 100kW turbines on Altamont's rolling hills. It not only operates its own wind farm but manufacturers its own successful turbine. U.S. Windpower also runs a number of 300kW prototypes.[33] The company originated in the early 1970s, when Stanley Charren and Russell Wolfe surmised that the future lay in understanding and attacking global energy problems. They needed to put their talents to work. At first they considered building electric cars, but a visit to engineer William Heronemus (see chapter 7) convinced them to change direction. Heronemus had a vision of a renewable energy future that was contagious, rhapsodizing about wind farms and floating windmills on the coast. Charren and Wolfe left the encounter determined to put some of Heronemus's visionary ideas to the test.

In 1974, the two entrepreneur-environmentalists incorporated U.S. Windpower. Before wind could be harnessed, money had to be raised. Within a few years they had commitments of $1.2 million from twenty-five investors. On New Hampshire's Crotched Mountain in 1978 they put up twenty 25kW windmills. It was the first wind farm in the United States.[34] Charren and Wolfe brought in talented engineers and crafts people to continue refining their product and perfecting their design. One reporter noted: "The goal has been to achieve perfection, testing one design, ripping it apart, and trying out another." Norman Moore, founder and head of Litton Industries microwave oven division, became president. Alvin Duskin, an entrepreneur in the garment industry, became the marketing expert.[35] It was Duskin who took U.S. Windpower to California, through his friend Huey Johnson, past president of the Nature Conservancy and at that time Governor Brown's chief of the Natural Resources Department. U.S. Windpower moved its headquarters and activities to Livermore, just west of Altamont Pass.[36] Between 1982 and 1987, the company enticed investments of more than $350 million in private funds to develop their Altamont projects.[37] The company has expanded to Carquinez Straits and is prospecting for new ventures in the Pacific Northwest, Montana, the Midwest, and even back in New England. Recent developments are documented in chapters 11 and 12.

In some ways, the early 1980s in California paralleled the gold rushes of the nineteenth century. But a more realistic comparison can be made with the boomtowns of the twentieth century. The small town of Tehachapi, for example, experienced all the positive and negative factors of a rapid influx of population. The town had originated as a railroad whistle stop, then people found work at a cement plant, in constructing the Los Angeles aqueduct (from Owens Valley to Los Angeles), and ranching. When the wind-energy boom hit, friction broke out between the established residents and the newcomers, who were young and often foreign. Paul

Gipe, then working as a publicist for the industry, described the town in the early 1980s as "America at its finest! Chaos!"[38] More Danish, German, and Italian speech was heard in town than English. Cowboys and Danes fought in the bars. Public hearings became shouting matches. Fraud and deception for a time became a way of life. The social discord and growth impact did not match that in the 1970s of Wyoming, but it was an exhilarating, maddening, five years. Gipe recalled: "You had to be there. It was like the Wild West"—but not the gunslinging West; Gipe had in mind the notion of community. Townsfolk felt they were "building something together—like the Amish at a barn raising. It was a heady experience."[39]

There was plenty to be done. In 1981, a year of dialogue between developer and investor, developers installed only 150 turbines, representing 10MW of capacity. But the years to follow were a frenzy of activity. In 1982 developers raised 1,200 turbines; the number jumped to 2,549 the following year and to the zenith in 1984—4,732 units. By 1985, construction was tailing off a little, with 3,922 turbines coming on-line. But half a decade into the 1980s, a portion of the California landscape had been transformed. Whirling windmills were at work at Altamont, Tehachapi, and San Gorgonio, with a total of 12,553 wind generators installed. Total capacity was 911MW, or 96 percent of the existing wind-energy capacity of the United States.[40]

1986: THE BUST

However, in the mining and energy business it is axiomatic that where there is a boom there is also a bust. The bust came in 1986. Federal tax incentives expired on December 31, 1985, attempts at saving the tax credits finding little sympathy in the White House. Affluent residents of Palm Springs with connections to the Reagan administration adamantly opposed any more tax credits for an industry they found offensive.[41] Furthermore, the price of oil was skidding downward, and with it went Washington's enthusiasm for renewable energy development. Wind skeptics, such as Rep. Pete Stark, whose district included Altamont Pass, publicly declared the wind business to be little more than a tax dodge.[42] Clearly, the industry's halcyon days had ended.

Major new companies such as Zond, Fayette, and FloWind found themselves on the brink of bankruptcy and Chapter 11 became part of the wind-energy business lexicon. Manufacturers were also hit and Danish companies faced retrenchment. Birger Madsen, a consultant representing the Danish wind-energy industry, stated in 1988 that the "Danish wind turbine manufacturers have been hard hit by the demise of the California market. The summer has been full of news of bankruptcies and financial difficulties within the industry. Naturally enough those hardest hit have been the firms most involved in the American market."[43] American manufacturers who produced marginal turbines went out of business. Even the historic Wincharger machine sank into oblivion: Energx Corporation, of St. Paul, Minnesota, produced its last batch of 450 Wincharger W200s in 1988, the only model surviving from the 1950s.[44]

Close on the heels of bankruptcies and cancellations came court cases. Cities sued federal agencies, investors sued wind-farm companies, wind-farm companies sued manufacturers, and manufacturers sued insurance companies. Perhaps the worst crisis occurred in early 1988 when Micon president Peder Morup confirmed that more than nine hundred Micon turbines had been brought to a halt for safety reasons. The problem was faulty blades. Repairs would cost millions of dollars. Who would pay for the replacements? Who would pay for the lost revenue (an average of $875,000 a month)? The insuring company, the Zurich Insurance Company, of Illinois, sued not only Micon but the Moerup Industrial Windmill Company, of Randers, Denmark, the Diversified Risk Insurance Brokers, and Nordic Special Risk Agency, Incorporated.[45] With such financial and legal maneuvering in progress, the emphasis on technology, reliability, and production seemed a secondary matter. The boom had become a bust. At the annual American Wind Energy Association—perhaps awkwardly scheduled to be held in fun-filled Hawaii—there was an aura of gloom. "The mood was uneasy and divisive and in sharp contrast to last year" reported *Windpower Monthly* journalist Ros Davidson.[46]

The gloom could not altogether undercut the accomplishment of the last few years, however. The bust was culling inferior machines and corrupt opportunists; leadership and integrity might win out. "We've possibly gone through the valley and better days are ahead," suggested Leonard Rogers, of the Department of Energy (DOE).[47] Certainly, progress had been made in determining a practical size for wind-energy installations. Most of the 12,553 turbines installed between 1981 and 1986 were of intermediate size, but a few small-turbine manufacturers had also come forward. Enertech, Jacobs, and Wind Power Systems, all U.S. companies, erected a number of 10kW to 40kW systems. They experienced difficulties and eventually all three dropped out of the business. This ended small turbine competition in the wind-energy grid-supply market. Small-scale wind turbines retreated to their traditional, limited market—the farm environment.[48]

At the other end of the spectrum, the failure of DOE research and development work on large-scale wind generators discouraged both buyers and builders. Companies such as Boeing, Hamilton Standard, Alcoa, Bendix, and General Electric withdrew from the market. Elsewhere, Denmark, Sweden, Germany, Canada, and the United Kingdom were experiencing far more failure than success. Proposals for development of large wind turbines—the best-known being a 100MW wind farm in Solano county, featuring fifty Boeing MOD 2—failed to attract investors and resulted in abandonment. The only such development occurred on the Hawaiian island of Oahu, where Hawaiian Electric Renewables Systems, Inc. installed an 8.4MW farm of fourteen Westinghouse wind turbines (600kW) and a Boeing MOD 5B, the last of the American giants.[49]

Most of the action focused on intermediate-size turbines—those ranging from 50kW to 500kW. Although investors rejected the huge turbines, since 1981 the average size of turbines has steadily increased, as engineers have gained experience in working with the technology. For instance, in 1981 and 1982, the number of turbines installed was divided equally between the 10–50kW category and the 51–100kW category. In 1983, 70 percent were in the 51–100kW range and 15 percent were in the 101–200kW range. There were also 10 turbines above 200kW.[50] The trend toward larger machines has continued. Most experts agree that in the United States the future average size will be 500kW. In Europe it will be near 1MW.[51]

DANISH DOMINANCE

As competition for supremacy heated up, the Danish design emerged triumphant. U.S. Windpower manufactured thousands of its 50kW to 100kW turbines, but they were installed only on Windpower wind farms (primarily at Altamont and Carquinez Straits). The Danish companies, exclusively producing three-bladed, upwind turbines, made quick inroads into the market and developers turned more and more toward Europe. In 1982, non–U.S. suppliers held 19 percent of the world market; by 1985, this figure had grown to 55 percent, the largest foreign supplier being the Danes. That year, Danish companies supplied more than 50 percent of the turbines installed in U.S. wind plants. The Danish government assisted in forming a number of joint ventures between U.S. developers and Danish manufacturers, these enterprises often being augmented by Danish financing. Zond Systems, for instance, installed and operated many (Denmark) Vestas machines in their Tehachapi wind farms; the SeaWest Energy group worked with Denmark's Micon at their 221-turbine Altamont Pass farm. FloWind, in addition to installing

Danwin turbines have been judged the most attractive Danish three-bladed turbines by many observers. (Author photo)

and maintaining their own vertical wind turbines, formed a joint venture with Denmark's Danwin.[52]

The dominance of the Danish machines dashed the hopes of some California wind-energy analysts. Part of the rationale for state tax credits and encouragement rested on the belief that the wind turbines would be manufactured within the state, thus creating jobs and revenue. With such strong aeronautical and defense industries located on the Pacific Slope, there was every reason for this assumption. However, it was the Danes, with their emphasis on low technology, use of lots of steel, and what researcher Matthias Heymann labeled "the craftsman tradition" that won the field. Vestas, Micon, Bonus, Windmatic, Nordtank, and Danwin—all Danish companies producing reliable wind turbines—became familiar names in California's wind-energy business.

Whereas U.S. companies lost out in the manufacturing competition, they did find a place in developing and managing the many wind stations. Some of the companies failed, quickly and miserably, but others were responsible, taking the lead in what was a new and novel undertaking. The importance of U.S. Windpower has already been indicated. Other companies, too, emerged. SeaWest Power Systems, Inc. has become a dominant force, presently operating some 2,252 turbines in Altamont, Tehachapi, and San Gorgonio passes (at Tehachapi, SeaWest operates 500 Mitsubishi turbines, the only Japanese entry in the market). Zond Systems, Inc. is another major producer at Tehachapi. They service and manage 2,600 turbines (260MW capacity). Wintec, Ltd., at San Gorgonio Pass, manages and services 700 turbines. FloWind, mentioned earlier, manages and maintains about 865 wind turbines at Altamont Pass and Tehachapi Pass. All told, about fifteen companies develop, install, manage, and maintain wind farms in the three complexes.

THE CONTRACT ISSUE

All the generating companies must sell their energy to either the PG&E or Southern California Edison. The Altamont harvest is collected at Tesla Substation, whence it enters the PG&E grid. The yield from both Tehachapi and San Gorgonio passes into the Edison grid. The two giant utilities contract with individual companies for purchase of the power production. Obviously, these contracts are crucial: there is only one purchaser. As outlined in PURPA, the utility companies must purchase this energy at the full "avoided cost" rate, but just what that figure should be has been the subject of vigorous debate. There are generally two major components to avoided cost: energy and capacity. *Energy* avoided costs are often called avoided *fuel* costs—the cost of the fuel that the utility does not burn because the wind farm is generating electricity. Energy avoided costs are understandable, but the *capacity value* is more perplexing. Generally, the capacity value reflects the *decrease in installed generating capacity* made possible because the wind plants exist. In California, capacity value is reasonably high because the wind arrays produce a significant proportion of their energy during periods of peak utility demand. This peak demand period is usually in the late afternoon of the summer months, when thousands of air conditioners are turned on. Fortuitously, this is also the time of day when the wind is flowing from the cool coastal regions to the hot valleys and deserts. "The coincidence of high wind farm output during periods of peak utility demand," wrote one independent analyst, "is responsible for the greater value of the wind farm avoided cost."[53] In 1983, this same analyst projected the avoided fuel cost for 1987 to be 7.55 cents per kWh. Added to this would be 1.18 cents for avoided capacity costs. He therefore judged the total avoided cost to be 8.73 cents per kWh.[54]

Determining the avoided cost to be paid to wind-energy companies is a subjective matter. As indicated in the above example, the analyst must guess the future price of energy (primarily from oil and natural gas) based on all kinds of economic, political, and environmental variables. It is an inexact science at best, voodoo at worst. In general, the prevailing view in the early 1980s, when most utility company officials signed ten-year contracts, suggested that about seven cents per kWh was a price that might allow for a profit if all went well.

Of course all has not gone well since 1983; at least not for wind energy. The price of oil has gone down, and therefore so have the avoided costs. "I don't see much future for wind if the cost of fossil fuels doesn't go up," stated Julian Ajello, a California utilities commissioner, in 1989. The low price of oil translated into an avoided cost of about three cents per kWh.[55] Selling their product at that price would soon lead the wind-energy companies to bankruptcy.

For the foreseeable future, the negotiation of new avoided costs agreements is the most important issue facing the wind-energy companies. It is often called "the eleventh-year issue."[56] Most wind-energy companies have thirty-year contracts with PG&E or SCE—the first ten years being on a set avoided-cost price, usually around seven cents per kWh. However, the last twenty years of the contract are on a floating scale, to be determined by new calculations of the avoided costs. Many of the contracts were entered into during the 1983 to 1988 period, thus, in the mid-1990s, negotiations were under way. Contracts are being negotiated with approximately three cents energy avoided cost and two cents capacity costs for a total of five cents. Many companies will need a higher rate to survive. Already, Fayette Energy Corporation and Zond Systems have shut down some turbines because they are losing money.

Stability is not within the lexicon of the wind energy world, but it must be noted that in the decade of the 1980s an idea finally became a reality. For all the unworkable engineering, state-sponsored hyperbole, shady business deals, tax avoidance, and just plain corruption, a new industry survived. Like most ventures it had its growing pains and in the late 1980s the idea of wind energy was associated with floundering technologies and failed investments. These criticisms and failures will be addressed in the next chapter, but it is well to remember that in California, messy though it was, a new industry was born. Those exciting years represented the living, spinning embodiment of the hopes of many Americans that renewable energy production would be more than mere talk.

CHAPTER 11

CATCHING THE BREATH OF THE SUN: PROBLEMS AND PROMISE

I plan to fly to the nation's capitol . . .
to do battle as Don Quixote did against windmills.

—SONNY BONO, MAYOR OF PALM SPRINGS, 1989

WHENEVER THERE HAS BEEN AN ECONOMIC BOOM IN THE AMERICAN West, throngs of human beings are drawn there. Gold miners, oil roughnecks, health seekers, water purveyors, uranium prospectors, land boomers—all have had their shot. Some made honest fortunes and contributed to the region. Others were notable for creative chicanery and outright corruption. The wind-energy boom of the early 1980s had both.

Reflecting on the early 1980s, Bill Vaughan, the legislative aide of U.S. Congressman Pete Stark, characterized the early industry as a mongrel: "A combination of environmental zealousness and the earth movement of the early 70s—the true believers—and the financial types who could have been selling shoes, I suppose. Weird marriages like that, you get weird puppies."[1]

The "weird" aspect of wind energy has spawned controversies, tragedies, and an occasional success story. This chapter will explore examples of the failures, the controversies, the triumphs, and the spread of the contemporary wind-energy business. I want to focus on the views of people who live in the towns or rural areas where wind-generating stations have been established, or have been proposed. People in such places have been explicit in their support or condemnation of wind-energy plants. The plants produced a necessary product; they also represented change to customary perceptions of landscape and lifestyle. The overarching theme of this chapter must be the recognition that no form of energy production is totally benign. Each will have an impact. The secret—and the challenge—is for society to choose the energy source which is sustainable, reasonable in cost, and the least offensive to both the environment and human perceptions.

Among the towns and cities most impacted by wind development in California are Palm Springs, Tehachapi, and Livermore. Outside the Golden State, the communities of Livingston, Montana, and Marshall, Minnesota, also yield specific insight into the problems and promise of wind energy.

The story of Palm Springs and the San Gorgonio wind generating stations provides a particularly interesting case study of the difficulties wind-energy proponents faced. As noted earlier, San Gorgonio Pass offered enormous potential: the massive San Jacinto and San Gorgonio peaks squeezed and contracted the eastward flowing wind, forcing it through the pass and into the Coachella Valley, a natural desert, now home to date trees, golf courses, swimming pools, second homes, and space. But after Dew Oliver's "blunderbuss" experiment, there was no further action until the late 1970s when the Southern California Edison Company (SCE) acknowledged the site's potential by establishing a wind-energy test center for trials of "innovative wind turbine designs."[2] The Los Angeles *Times* greeted the news enthusiastically, running an editorial entitled "Congratulations to a Maverick."[3]

Between 1980 and 1985, the utility company tested five turbines: a 1.3MW Bendix three-bladed, horizontal machine; two vertical-axis turbines (a 50kW and a 500kW, both manufactured by Indal, a Canadian company); a 100kW Wenco, two-bladed, horizontal-axis turbine; and a 330kW Howden, three-bladed, horizontal-axis turbine from Scotland. SCE let it be known they were testing these turbines to collect data "to support decisions relative to their [the turbines] large-scale use."[4]

Considerable publicity accompanied the erection of the turbines. Governor Brown could not attend the dedication of the 195-foot Bendix, but Russell Schweichart, a former astronaut who was now chairman of the CEC, lauded the efforts of the utility company. Schweichart likened wind-energy production to "catching the breath of the sun."[5] After adjustments, each turbine performed reasonably well, but neither SCE nor wind-energy financial investors were inclined to give financial backing for large-scale production and installation of any of these prototypes. These results were not encouraging; nevertheless, the economic incentives of the early 1980s lured many players into the deserts of the pass.

Not only were the tax inducements attractive, many of the best sites were located on federal land. The Bureau of Reclamation, manager of these areas, leased out land at bargain prices, convinced that promotion of wind energy helped fulfill the agency's 1976 commitment to conservation and wise land use. All this was well and good, but it resulted in a "wind rush." Some companies, indifferent to quality and energy production and mesmerized by the financial stakes, put up towers and turbines fast, with little regard to quality. Bill Adams, a reputable wind plant developer, later summarized the mood: "A lot of schlocks were getting into the business and selling prototype machines on a mass basis that didn't work."[6] This led to disillusionment in Palm Springs and other neighboring communities. The wind-energy industry's credibility was getting tarred.

International Dynergy, Inc., of Palm Springs, provides an example of this type of irresponsible activity. Dynergy put up two wind-turbine arrays, one named Cabazon, the other Maeva. Problems soon developed. On the financial side, the Bank of America severed its relationship in November 1985, suffering $1.5 million in unsettled debts. The company had also played fast and loose with its limited partnership investors, selling more than three hundred wind generators to investors in the Cabazon site by the end of 1985, but erecting only about half that number. Moreover, the turbines in place proved unreliable. At Cabazon, even at the best of times, only sixty-two turbines—fewer than half—were in operation, and that figure sometimes slipped as low as thirty-one. At the Maeva site, the figures were just as dismal. In the last quarter of 1985, Maeva produced a pathetic 377,750 kWh, approximately 12 percent of projected production.[7]

The company's miserable performance resonated far beyond mere statistics. Built close to I-10, Cabazon's failure did not go unnoticed by the thousands of travelers using the highway. The derelict site, with its downed turbines, broken blades, and general technological chaos, magnified perceptions of the industry's problems. Cabezon became, in the eyes of one reporter, an eyesore of broken and twisted blades: more a war zone than a wind park.[8] The Cabezon and Maeva situation also attracted the attention of the Internal Revenue Service. In the pragmatic view of the IRS, if investors received tax credits, the wind turbines should perform.[9] The IRS sent investigators to Palm Springs and the company went still closer to disaster when employees appeared to be attaching old helicopter blades to defunct units in anticipation of the IRS inspection.[10]

The company's problems were compounded when two employees fell sixty feet from a wind tower after a blade severed the basket cable. The broken bones of one worker, Mario Botello, mended, but Andreas Romero, the other victim, sustained brain damage and could neither walk nor talk. Lawyers filed a legal suit demanding millions of dollars, only to find that Dynergy had stopped payment on its insurance premiums well before the accident.[11]

More law suits were on the way. Normally passive limited partners—119 of them—filed a lawsuit against Dynergy for $1.9 million, claiming that they had been deceived in investing between $120,000 and $140,000 per wind turbine with no return.[12] Now the IRS threatened the cherished tax breaks by forming a nationwide team to investigate the marketing and operation of wind-energy firms.[13] Meanwhile, the Bureau of Land Management, federal landlord of the two parks, ordered that the eyesore of broken and twisted blades be removed from Cabazon. International Dynergy and its subsidiary Transworld Wind Corporation, in a hasty retreat, countered this move by turning over management of the turbines to three hundred investor-owners, thus creating a bureaucrat's nightmare of multiparty ownership. Lantson Eldred, the investors' lawyer and spokesman, spoke confidently of reform and rebirth. He rather mockingly paraphrased Mark Twain's "rumors of our death are greatly exaggerated"[14] and protested the BLM edict, claiming that the machines could be repaired and thus should not be removed. In a prophetic statement regarding not only the blades but the whole Cabazon fiasco, Eldred remarked: "I feel like I am putting Humpty Dumpty back together again."[15]

While the lawyers and bureaucrats were struggling over the fate of Dynergy, reputable companies faced unanticipated mechanical predicaments in the desert environment. Desert winds produce dust storms that can take the paint off an automobile or pock a window to the point that it is no longer translucent. As wearers of contact lenses well know, a dust storm can cause severe tearing. The wind machinery experienced related problems. Dave Kelly, field manager at SeaWest Tehachapi, has seen particles the size of small rocks lodged in nacelles high off the ground, blown there by wind. Bearings and machinery became clogged with dust.[16] Such environmental and mechanical challenges cannot be resolved in the laboratory.

Another unforeseen difficulty appeared in late 1987: Danish fiberglass windmill blades developed structural problems—what might be called fatigue cracks. By February 1988, some 20 percent of the Danish machines had to shut down. SeaWest Energy Company, one of the most responsible companies operating in San Gorgonio Pass, suffered the worst, with 494 of their 870 units offline. Paul Gipe put a hopeful spin on a bad situation by announcing that now companies could retrofit the errant machines with more efficient blades, but nothing could alter the fact of massive failure. With hundreds of turbines at a standstill, it was visually evident.[17]

Since the introduction of wind plants in 1981, some Palm Springs residents had expressed reservations about them. They often were openly hostile toward the intrusion, and as the wind-farm failures increased, so did the number of opponents. By 1985, Palm Springs Mayor Frank Bogert felt comfortable in voicing his opposition. The ranks of industrial metal stalks and bristling blades that materialized near the highway entrance to the town detracted from the charm of the place and threatened economic consequences. "We don't think tourism and industry go together," said Mayor Bogert, "and all these windmills look like industry." In March 1985, the city of Palm Springs filed suit in the U.S. district court in Los Angeles against the Bureau of Land Management and seven wind-farm developers, seeking to force them to dismantle, move, or camouflage some of the existing units and to stop them from erecting others until stricter standards were agreed upon.[18]

The rapid proliferation of wind turbines, the high percentage of failure, the evaporation of the oil crisis, and the shady business deals of some developers further hardened the oppositional attitude of many Palm Springs residents. In May 1987, the Palm Springs *Desert Sun* ran an editorial, "Time to Get Tough on Windmills," proclaiming that with the oil glut the wind generators were no longer necessary. The town, however, was "stuck with the turbine forests which have spoiled the scenic beauty of the northern end of the Coachella Valley." The editorial recommended that Riverside County should take a hard look at the operators and only those "who can

Long lines of wind generators such as these Mitsubishi turbines at Tehachapi, California, have inspired military metaphors. (Author photo)

prove the financial and moral commitment to finding long-term solutions to U.S. energy problems should be allowed to continue operation."

The zenith of the opposition to the four thousand Palm Springs wind turbines came when Sonny Bono became mayor. A high-profile Hollywood star, who had achieved a certain fame as the husband and singer-comedian partner of Cher Bono, the new mayor commanded media attention.[19] Although not particularly articulate, Bono could transform a local story into a national event. He also found considerable backing among the rich and famous of Palm

Springs. In February 1989, with a supportive city council, Bono vigorously opposed the proposed construction of seventy-four new 500kW turbines atop Whitewater Hill in the San Gorgonio Pass. In a letter to the *Desert Sun*, one opponent of the windmills said they scarred the desert and soaring mountain landscape. The scene made her think of "someone driving big railroad spikes into a beautiful painting by Picasso or Rembrandt."[20] Others joined Bono's campaign on the basis of noise pollution. Some residents with homes within a two-miles radius the wind turbines found the *whooshing* noise of the turbines intrusive, especially at night, when the sound was intensified by its bouncing off the atmosphere: the specific sound might differ, depending on the size and the design of the turbine. One observer had earlier described it as "a pulsing beat . . . [like] a heartbeat magnified through a powerful speaker system."[21] Such a reminder of mortality was for some people surely a cause for sleeplessness.

With support from a vocal city councilman, Bill Foster, who claimed the turbines were "as damaging to Palm Springs visually as strip mining has been to towns and villages in Kentucky and West Virginia," in June 1989 Mayor Bono decided on a dramatic move. Calling the turbines ugly and "a tax write-off for the owners," Bono announced that he would "fly to the nation's capitol . . . to do battle as Don Quixote did against windmills."[22] The announcement proved to be more dramatic than the battle. The Bono group met with the secretary of the interior, Manuel Lujan, who with diplomatic aplomb assured the mayor he would look into the situation.

The mayor, in fact, represented some of the people but not all of the people. A study by academic investigators Martin Pasqualetti and Edgar Butler revealed a divided attitude among the residents of Palm Springs. People who had built homes in the outlying, isolated areas of Desert Hot Springs, North Palm Springs, and Whitewater had done so in expectation of solitude. For their part, the invasion of the wind turbines was understandably not appreciated.

However, a telephone survey of 165 Palm Springs residences, with an error factor of plus or minus 7 percent, "demonstrated that the vocal opposition to the wind turbines so commonly expressed is not borne out in fact."[23] The survey found that 51 percent of those polled considered the development of wind energy to be a good thing, 21 percent were opposed, 23 percent were neutral, and 5 percent did not respond. Only 22 percent believed that the wind turbines should be dismantled; almost 50 percent thought they should not. The majority of those surveyed believed that the wind farms were a good use for undervalued land. Those who strongly opposed the turbines and the noise intrusion often lived within two miles of the turbines. The findings contradicted "the negative impression which one gleans from talking with local legislators and residents."[24]

In a major metamorphosis, Mayor Bono reversed his position in August 1990.[25] The wind turbines were no longer the enemy but the city's friend, perhaps even savior. The city was strapped for money and Bono was loath to increase local property taxes. He now listened to such wind-energy leaders as Clare Lees (Field Service and Maintenance Company), Fred Noble (Wintec, Ltd.), and Curt Maloy, president of the new Desert Wind Energy Association.[26] A full-page Wintec advertisement appeared in the *Desert Sun*, featuring a photograph of a wind turbine that will "power 33 average homes . . . avoids burning more than 320 barrels of oil each year . . . [and] is quiet and reliable." The ad did not fail to point out that the turbine "provides jobs and pays taxes."[27] Elaborating on the advertisement, Clare Lees argued in June 1989 that critics ought to look at the economic benefits of this local industry, which included a payroll of over $6 million per year and $4 million in supplies purchased. Furthermore, the companies paid taxes on $550 million of assessed valuation, equal to four thousand homes. "Yet the windmills require no schools, sewers, water service, or paved roads," wrote Lees, "and very limited police and fire protection." In a direct appeal to Bono, Lees added: "Sonny, don't pout because the windmills won't go away. Use them to the city's advantage, and to the advantage of your constituents, the people of Palm Springs."[28] Some residents now also gave a different view on the issue of tourism. The wind farms, they said, were not a detriment but an asset. Gary Carpenter, a resident of the Morongo Valley, noted that "on any given day you can see cars or tour buses parked near one of the wind farms." Far from wearing looks of disgust, these tourists with camera in hand wanted to know more about these spinning curiosities. The wind farms had become another Palm Springs tourist attraction. Carpenter suggested that the city might "publish brochures about the windmills so the tourists don't have to scrounge for the information."[29] By 1991, a Palm Springs columnist was writing: "Like it or not, the windmills have become a tourist draw."[30]

By July 1990, Mayor Bono fully appreciated the economic arguments. Admitting that he had been "pretty boisterous" against wind energy, he now recommended that the city look for ways to annex some of the wind-farm area. "If we annex the property, the city would receive $1.6 million [annually] in property taxes," he figured. Sales tax on the produced electricity would also be available. By early 1992, the cities of Palm Springs and neighboring Desert Hot Springs were locked in a turf war over annexation of wind-plant land. The city that got the most would see the greatest increase in its tax base and sphere of influence.[31]

By 1990 politicians, residents and the wind industry of Palm Springs had resolved their most serious differences. When Bono reversed his position, Fred Noble of Wintec, Ltd., announced "This is a new beginning."[32] And indeed, so it has proved to be. The American Wind Energy Association lifted its ban on conventions at Palm Springs and in 1992 the association held a remarkably successful gathering in the city. The Desert Wind Energy Association, under the leadership of Curt Maloy, has taken a leadership role in the city, sponsoring local events, including a Windfair on Earth Day.[33] The association sponsors educational videos, arranges tours, and is pushing for construction of a highway pullout, where tourists would be able to view the turbines. Maloy believes the wind-energy companies have evolved from pariahs in 1985 to good neighbors in 1992.[34] Paul Gipe characteristically commented in 1989 that the trade "should thank Sonny Bono. Because of his criticism, he's made our industry newsworthy again and given us an opportunity to tell our story."[35]

However, the wind-energy producers at San Gorgonio do not feel secure. Although the wounds are healing between city and industry, the problem in the mid-1990s is the price of electricity. As mentioned in chapter 10, most wind-electric producers have thirty-year SO4 (Standard Offer 4) contracts with the local utility company, specifying an accelerating price in the first ten years. The price per kilowatt-hour for the final twenty years of the contract must be negotiated and that price will be crucial to the survival of the industry.

In 1990, Ron Luxa, supervisor of power contracts for SCE, and Charles Imbrecht, chairman of the California Energy Commission (CEC), squared off on the coming controversy. In an interview, Luxa pointed out that negotiation of the contracts was based on the 1982 price of oil—$30 per barrel—and that price had dropped from one-third to one-half. "Edison pays only about 3 cents per kilowatt hour for electricity generated by gas and oil-fired plants," the SCE representative was quoted as saying, "but they still are required to pay most wind developers an average of 7 cents a kilowatt hour." Luxa figured that in 1989 SCE paid $40 million more than if the company had produced this energy from oil and gas-fired plants. This additional cost—about $10.14 per customer per year—was passed on to ratepayers. As quoted in a *Desert Sun* article, Luxa admitted that wind energy was probably here to stay, but that in the future "they will have to compete with other resources, and that is the criteria that their success will depend upon." In the view of the utility company, the rate in the new contracts would certainly be cut dramatically. But from the point of view of the wind operators, that would be disastrous.[36]

Countering, in the same article, Charles Imbrecht stressed that the San Gorgonio wind-plant operators "have not had a level playing field in the past." The problem with Luxa's logic, he said, was that it did not reflect the true cost of electrical generation. To determine that true cost (to society, if not directly to SCE) the utility company would have to take into account the negative impacts of fossil fuel plants on the environment—such as air pollution, global warming, and water consumption. Furthermore, said Imbrecht, the environmental cost of transportation of fuels and disposal of waste must be factored in. In the long run, he claimed, new technologies such as wind energy would be attractive to consumers: "Figuring in those [social] costs, the state's energy commission found wind energy to be the least expensive electrical source in the state."[37] Imbrecht gave no specific cost per kilowatt-hour, but it was clear that when contract negotiations took place between utility company and producer, wind-energy companies would have a powerful friend.

The issue is crucial at San Gorgonio. In late 1992, when companies such as SeaWest were making money and plowing profits into retrofitting and replacing old machines, they were also preparing for a time when "the dime may be cut to a nickel."[38] In early 1995, the outcome of contract renegotiation was still open. The final arbitrator in determining a price per kilowatt-hour that is fair to both utility customers and the wind-energy companies was expected to be the California Public Utility Commission. SCE engineer Robert Scheffler, who is knowledgeable regarding both SCE and the local wind-energy companies, believed decisions hinge on "what the public utility commission finds reasonable."[39]

TEHACHAPI

One hundred and fifty miles to the north, the little town of Tehachapi accepted the wind turbine invasion much more gracefully than did Palm Springs; not, however, without controversy. Tehachapi had none of the size, wealth, or glitz of Palm Springs. An offspring of railroad construction, the town found its livelihood in railroad activity on the pass, the building of the massive aqueduct from the Owens Valley to LA, the existence of a state prison, and traditional livestock ranching. Over the years, growth had been zero to slow. Thus, the turbine building of the early 1980s was observed by local people with a hopeful but cautious eye. Almost overnight, Tehachapi transformed into a boomtown, buzzing with foreign workers, different customs, strange languages. Danes, Germans, Italians flocked in.

Over the last decade, the residents of Tehachapi have slowly made their peace with the turbines. The community has come to recognize that "they are part of a significant experiment in creating a new future" and the Kern County Wind Association has fostered this feeling through community events. A notable celebration was the 1991 Wind Fair. Over a period of two days, a crowd of fourteen thousand people focused on the wind. They bought T-shirts (For a Breath of Fresh Air—Tehachapi), flew kites, watched sky divers, rode bicycles among the wind machines, and listened to music.[40] The town has learned to live with this new industry, partly because it boosted the economy, partly because residents realized that visual disturbance was a small price to pay for energy.

Zond Systems, Inc., emerged as the dominant company at Tehachapi Pass, contracting to provide wind energy power to SCE. The Victory Garden wind farm was started in 1981. It was considered to have reached "build out" when 1,338 turbines (aggregate capacity, 120MW) had been installed. The first units placed at the site were rated at only 40kW; the newer turbines are larger, most rated at 225kW. This evolution from modest size to increasingly powerful turbines is typical throughout the three major wind-farm areas. In 1990 and 1991, Zond substantially boosted its Tehachapi operation, adding 342 Vestas 225kW turbines at its Sky River station, located fourteen miles northeast of the Victory Garden facility. With smaller installations at both San Gorgonio and Altamont, Zond produces about 15 percent of California's wind energy.[41]

FOR THE BIRDS

Our birds would look like they went through a Cuisinart.

—RUSSELL HEMSATH, RACING PIGEON ENTHUSIAST

The Zond story has not all been about success. The company was roundly routed in its attempt to place a desert wind farm near Lancaster, California and by 1986 the threat of bankruptcy haunted the company. Although the company fought its way back from this debacle, it ran into further trouble at Tejon Pass, the major thoroughfare between the San Joaquin Valley and the Los Angeles Basin. After measuring the wind resource for five years, in January 1987 Zond announced an agreement to lease 270 acres of windswept land owned by Ron Ralphs, whose father Oscar had established the spread, near Gorman, in 1881. Zond proposed to erect 458 turbines, building an array of towers 70 to 150 feet high.[42] It was to be the first wind-energy station in Los Angeles County.

Although LA County was widely recognized as one of the worst polluted regions of the nation, the prospect of energy-without-pollution had little impact on the county's many residents. However, the expectation of spinning turbines did. Opponents, backed by the powerful 200,000-acre Tejon Ranch, formed a committe, the Save the Mountain Committee, to fight the project. Zond president Jim Dehlsen attempted to mollify opposition by proposing to establish a fund that would assist local community projects. Dubbed the Meritus Fund, it could have amounted to more than $160,000 a year, depending on the number of wind turbines and their performance. Tom Gray, executive director of the American Wind Energy Association, expressed cautious support for the idea, noting that it "may be a good way to compensate for the perceived impacts on local communities." Gray did not want to add to the cost of wind energy, but he was quoted as saying: "If this is necessary to sell people on wind energy, then it needs to be done."[43]

Opponents were unimpressed. Although a few Gorman residents found the offer attractive, most agreed with Tejon Ranch counsel Allene Zanger that the proposition was just a little short of bribery. Even though the project was environmentally sound and would produce clean electric energy for forty thousand homes, the cards seemed to be stacked against the wind plant.[44] Susan Hulsizer, active in the Sierra Club both locally and at state level, led an energetic campaign for the Save the Mountain Committee. She enlisted not only the Sierra Club but also the Audubon Society. Dehlsen found support at World Watch, the Rocky Mountain Institute, Zero Population Growth, and Ralph Nader's group—all organizations that were against the growing threat of global warming and air pollution.[45] Opponents claimed that the visual blight would ruin "the freeway experience" through Tejon Pass, as well as a favorite wildflower area. Dehlsen and Paul Gipe countered, noting that only about 5 percent of the turbines would be visible from interstate 5 and that the real opposition came from developers who wanted no intrusion on future real estate projects.[46]

Unfortunately for Zond, they had located the proposed wind farm in California condor habitat, or what would be condor habitat when the birds were released. State, federal, and private wildlife experts had invested countless time and effort in a bid to save the condor and any threat, no matter how slight, was considered intolerable. Zond promised continually to monitor the radio signals from the birds and shut down the wind turbines when condors entered the vicinity. Linda Blum, of the Audubon Society, was not convinced, noting that monitoring "tended to be fairly unreliable" in the past.[47]

On August 16, 1989, the Los Angeles County Regional Planning Commission held a hearing on the issue. The Save the Mountain Committee submitted a petition of three thousand signatures against the project and a vocal audience of more than two hundred people packed the meeting room. A last-minute pruning of the project from 458 turbines to 319 did little to mollify the opponents. Sherry Teresa, a wildlife biologist for the California Department of Game and Fish, noted the danger to traditional condor flyways and also pointed out deficiencies in Zond's environmental impact statement: bald eagles, peregrine falcons, and Swainson's hawks, all on the endangered species list, had been overlooked. Another avian argument came from Russell Hemsath. Speaking for the California State Racing Pigeon Organization, Hemsath graphically objected that, if the project went ahead, "our birds would look like they went through a Cuisinart." The image of chopped up pigeons and raptors, executed by turbine blades, was telling. The planning commission unanimously rejected the Zond proposal and the Tejon proposal died.[48]

THE DILEMMA

Whenever we think we've found the perfect
environmentally benign form of energy production
we find that there's a problem.

—DAVID NESMITH, SIERRA CLUB, 1992

The Zond argument, stressing the national and global environmental position (global warming, air pollution), had proved to be no match for organized, well-funded, concerted local action. As two analysts put it retrospectively, "The macro-scale benefits of wind energy . . . seemed to make no impression and failed to counteract the micro-scale, local concerns of opponents."[49] But the positions taken by local people and environmental groups came in for criticism. Ken Karas, Zond chief executive officer, found it ironic that "a county with some of the worst smog in the country [rejected] a power plant that could help reduce dirty air."[50] And in a hard-hitting column, environmental journalist Alston Chase castigated the local NIMBY ("Not in My Backyard") reflex "dressed in the language of ecology." The forthright Chase concluded: "The spirit of John Muir, Teddy Roosevelt and Rachel Carson is being co-opted by affluent practitioners of primitive chic more concerned with property values than with ecological sustainability."[51]

This phenomenon of local protest halting wind energy development was repeated at Cordelia Hills, a small town in northern California. Behind the town—just off I-680, near Benecia—rose the Cordelia Hills, where wind speed averaged 16 to 18mph. The CEC had scheduled the hills for wind development, but when wind-industry developers prepared a draft environmental statement and the Solano County Planning Commission and county supervisors held hearings in 1987, opposition emerged. Residents did not want wind turbines to be visible from their homes and they found support among expanding real estate interests in nearby Sky Valley. In a split vote, the supervisors rejected the development. One supervisor cogently expressed the dilemma faced by the wind industry: "If you're a cattleman, you call sheep unsightly. If you're a homeowner, you call wind farms unsightly. If you are a rancher, you call housing developments unsightly. It all depends on whose ox is being gored. How can we tell some people they can build unsightly houses and tell another they can't build unsightly wind turbines?"[52]

By the late 1980s, the evidence indicated that large-scale wind installations would not be permitted in regions close to housing or in areas scheduled for development. As discussed below, many people who do not have a negative attitude toward wind energy in general, do if sites are to be placed within five miles of their homes. This perception became clear in 1989 when U.S. Windpower had little difficulty in obtaining a permit to install four hundred wind turbines in the Montezuma Hills, west of Rio Vista, at the western end of the agriculturally rich Delta region. The Solano County site featured not only an outstanding wind resource, but remoteness. The hills were at least fifteen miles from the nearest interstate freeways, I-680 and I-80, and the city of Fairfield. Opposition did emerge, but the site was "off the cognitive map" of most county residents and therefore there was a "lack of a critical mass of a nearby, affected public."[53] The Montezuma Hills project sailed through hearings held by the county planning commission and the county supervisors and, with 617 wind turbines, is now fully installed. It is the fourth biggest wind farm in California.[54]

These three examples, Tejon Pass, the Cordelia Hills, and the Montezuma Hills, confirm the ambivalence of Californians toward wind energy. They embrace the notion of clean energy, but they were not convinced that specific wind farms should be supported. It was a matter of trade-offs. Landscape design expert Robert Thayer has suggested the hypotheses that people from rural backgrounds are more accepting of wind-energy development than people with urban backgrounds.[55] This certainly seemed to be true of the Altamont Hills, east of the suburban town of Livermore. For ranchers, the trade-off has been good financially. In 1983, Joe Jess Sr., for instance, had hundreds of wind turbines sprouting on his six hundred acres. In a newspaper interview, he looked forward to more; by his calculations he would soon be a millionaire, concluding that this latest endeavor was about "twenty times better than ranching." Virtually all of the other twenty ranchers of the Altamont Hills had followed Jess's example "inspired by ceaseless winds just like the forty-niners were crazed by gold nuggets."[56] Jess's son Joe Jr.

worked as foreman of a turbine construction crew, earning good money during a time when the cattle market was sagging. One young rancher stated: "I get along great with the cows, but I'm getting along with the windmills just as well." Landowners received a set fee for each turbine installed and a percentage of the price paid for the energy sold: it is little wonder they embraced development. Land with wind-development potential increased in value from around $400 per acre in 1980 to $2,000 per acre in 1987.[57] Moreover, more than 90 percent of the land is still available for grazing.

But Altamont residents who had no financial interest in the venture did not share the ranchers' enthusiasm. They hated the noise, particularly the ESI model (Energy Sciences, Inc.), which made "a high-pitched aerodynamic whizzing sound." The 400Hz gear-engagement frequency was "broadcast by the blades." Compounding these disturbances, the rotor created a thumping, helicopter-type noise each time a blade went behind the tower, a common problem with downwind turbines.[58] One couple, owners of a ranchette on the eastern side of the hills, protested so vigorously about the ESI machines that they were able to sell their house to the offending wind farm, which then used it as its headquarters.[59]

Two residents of Dyer Road on the west side of the hills have long fought the windmills. Darryl Mueller and John Soares complained about turbines operated by U.S. Windpower. In a 1991 newspaper interview, Mueller protested that the turbines were "within three-quarters of a mile of my house and I can hear these windmills beating against my windows late at night." Soares said of the night-noise level: "Sometimes it's almost unbearable." The two neighbors hired James Chester, of EnviroNoise, to see if the decibels consistently exceeded the fifty-five decibel limit established under U.S. Windpower's 1981 permit.[60] Measuring the sound at four feet off the ground, Chester found the company consistently in violation. But when U.S. Windpower hired its own consultant, the sound was measured at ground level. This suggested that the

This aerial view of wind turbines at Altamont Pass illustrates the compatibility of wind-energy production and traditional use of the land. Approximately 90 percent of Altamont land is still available for cattle grazing. (Courtesy WindMaster U.S.A, Inc., Byron, Calif.)

turbine noise was consistently below fifty-five decibels. Chester argued that by testing at ground level the company used a "muffled microphone," but given a standoff of experts, U.S. Windpower carried the day. Mueller and Soares have vowed to continue the fight, Mueller organizing a Save the Eagles campaign.[61]

Over the years, opponents of the Altamont wind farms have been especially effective in enumerating the number of failed units. In 1984, Sylvia White, a professor of urban and regional planning at California State Polytechnic University, Pomona, vented her disgust by writing that "the unsuccessful firms will leave us with their wreckage—those ugly, useless monstrosities that distort what were once some beautiful scenic areas."[62] In that same year, the Alameda County Board of Supervisors denied permits for new windmills on the basis of negligence in repairing or removing old ones. Approximately 557 older-model turbines had stopped working and critics of the wind companies urged the supervisors to refuse new permits. They did.[63]

Perhaps more perilous for the wind companies was the fact that broken and stilled units undercut the public's confidence in the wind experiment. A representative of Teamsters Union Local 70 called the wind farms a "scam on the taxpayers."[64] In March 1983, Congressman Stark, whose district, as noted earlier, included the Altamont Hills, announced his opposition to further federal tax credits. If tax credits had any chance of extension beyond the end of 1985, that hope died with Stark's opposition. Much of his disillusionment stemmed from the unreliability of the units. In the congressman's view, "with what taxpayers are paying, we'd be better off hiring thousands of kids to sit on bicycles and pedal away to produce power for our toasters. It would be cheaper."[65]

In retrospect it can be seen that legislators did not err in providing federal and state tax credits for the wind companies of California. Without such incentives, nothing would have happened. However, lawmakers should have tied tax credits to energy *production*. It

Travelers on Interstate 580 at Altamont Pass, California, encounter an awesome number of wind generators—considerably more than six thousand of them. This string of early turbines may have been the ones to inspire Sylvia White, Cal Poly Professor of urban and regional planning, to dub the Altamont windmills "exoskeletal outer-space creations." (Author photo)

should have been a performance-based plan. Instead, in an egregious lapse that seriously damaged the reputation of a struggling industry, legislators tied tax credits to *investment*, or what experts termed a capacity-based formula. By 1985, Carl Blumstein, a wind-energy expert with the University of California Energy Research Group, and others exposed the folly of that policy. In effect, whether the turbine worked or not made little difference to the manufacturer, developer, or investor: what counted was the rated capacity. Blumstein asked the state to change "to performance-based subsidies *as soon as possible*."[66] Almost every responsible analyst and wind-company executive recognized that awarding tax credits on the investment basis alone invited poor machinery, attracting swindlers who profited from an atmosphere "ripe with fraudulent activity."[67]

However, much of the decade-long protest over Altamont's 6,500 wind generators had little to do with economics or reliability. As at Tejon Pass, it was environmental groups who were most vehement in opposition. One such outspoken group was People For Open Space/Greenbelt Congress, an organization committed to preserving a swath of open space to the east of the San Francisco Bay Area. People For Open Space opposed the spread of suburban housing and the rapid growth of small towns, wanting to preserve agriculture, wildlife, and watershed areas. Although the Altamont wind farms helped achieve all of these objectives, the organization opposed wind-farm development. In fact, Mark Evanoff, director of the group, treated the wind turbines as if they were as onerous as a nuclear power plant, proclaiming that "we eventually will have to decommission the windmills."[68] For People For Open Space and its Sierra Club and Audubon Society supporters, wind turbines were industrial culprits, imposing on a pristine environment. "The greenbelt is not the place for light industry," stated Evanoff, refering to the Altamont wind station.[69]

Many people agreed with Evanoff's assessment. Sylvia White described the wind turbines as "exoskeletal outer-space creations"

As a result of a growing population and industrialization, California has been transformed from a land of promise to a landscape of compromise. Photo by the author shows California aqueduct, wind generators, and high-tension transmission lines.

that were "eerie to behold; towering, stilty-looking, at once gawky, yet graceful." The turbines, said White, who was not afraid of sounding almost bizarre, possessed grotesquely anthropomorphic characteristics—"long, sweeping blades attached to what ought to be their noses" with "legs . . . frozen in concrete, stationary but seemingly kinetic." For White the "once-friendly pastoral scenes now bristle with iron forests."[70]

Although the wind turbines occupied only 5 percent of the land, thus preserving agriculture, watershed, and wildlife (except perhaps for raptors), the expansive Altamont site did not measure up to the public's perception of what a natural greenbelt should be like. Evanoff and White highlighted public perceptions that could not be denied. They underscored landscape realities that could not be camouflaged. The immediate area of the PG&E Tesla Substation was (and is) a maze of spinning turbines, power poles, overhead wires, cyclone fences, and barbed wire—all part of the requirements of electrical power production and transfer. To the east of the substation is flowing water—but not the kind to encourage riparian plants or wildlife. This is the California Aqueduct, a sterile, cement canal. This, too, is an artificial, commercial manipulation of nature. The hand of humans has been placed heavily on this land.

The Altamont wind-farm operators, however, did not perceive themselves as industrialists destined to be in a continual fight with environmentalists. Patrick Agnello, who worked for Howden Wind Parks and who in his university days was an antinuclear activist, expressed astonishment at the wind-farm protest. Agnello recalled: "Wind energy used to be the darling of the environmental movement."[71] Indeed it did, but many environmentalists either did not understand or did not wish to recognide that no power production source is totally benign.

The friction at Altamont highlighted historic, ongoing differences between utilitarian conservationists and the purists or preservationists. One side stood for safe, nonpolluting, renewable energy

The spread of wind farms has faced determined opposition by environmental groups. A primary argument is that generating stations industrialize the landscape as is clearly evident in this photo taken by the author near the PG&E Tesla Substation at Altamont Pass, California.

development; the other argued for scenic preservation. The latter position wanted no new energy development anywhere, said it was unnecessary, and relied on consumer conservation to meet future energy needs. Wind-energy proponents believed that total reliance on conservation was unrealistic idealism. Even if demand were to remain steady, energy production units of all types—nuclear, oil, gas, and wind—wear out. Sooner or later they must be replaced. And when the positive and negative factors are compared, wind-energy proponents believe that their industry comes out of the contest favorably.

Some people find wind generators far from ugly. Jeff Greenwald, a freelance writer, wrote of Altamont Pass: "They are not ugly, these wind turbines, bristling on the green crest of the Alameda hillside like a sparse mohawk. . . . One can almost mistake them for kinetic sculpture. Cows and horses graze among them; hawks wheel overhead. It's a far cry from what goes on around the business end of an oil refinery or a nuclear power plant." The turning blades of the Fayette wind farm moved Greenwald to comment: "There is something deeply satisfying about watching power being produced in such a clean, elemental way."[72] Can one, observing the windmills, combine form with function and see art—i.e., "kinetic sculpture"? To do so is to progress toward reconciliation between technology, earth, and art. To the south of Altamont, at Tehachapi, Jerrie Cown, director of the Tehachapi Museum, seems to have attained that reconciliation. Although she would prefer not to see turbines in a few areas, in other places she has fondly accepted this addition to her visual landscape. "In the evening, they sparkle." she told a newspaper. "They're beautiful. They have a pink glow at sunset. I love to hike up around them."[73]

Such a reconciliation—central to the acceptance of wind turbines—has received much attention from Robert Thayer, an environmental design specialist at the University of California, Davis. Thayer has studied society's perceptions of the wind farms in sophisticated cultural terms. For him, most Americans have been heavily influenced by the interplay between *topophilia* (love of land and the pastoralist lifestyle) and *technophilia-technophobia* (a love-hate relationship with technology).[74] The wind turbines can produce "landscape guilt," because they violate our historic perception of what landscape ought to be. Leo Marx's *Machine in the Garden*, although providing no insights specifically regarding wind turbines, creditably documents and interprets the historic dilemma of love for pastoral nature and attraction to technology. The paradox still operates in the Altamont Hills.[75]

In 1986, Thayer's Center for Design Research sent out six hundred questionnaires, asking recipients to respond to six photographs of the Altamont Hills wind-turbine arrays. Half of those surveyed lived within ten miles of the Altamont wind farms; the other half lived approximately fifty miles away. Thayer received two hundred valid responses. In regard to the results, Thayer and coauthor Carla Freeman said in their report that "the subjects as a whole had mixed opinions about wind energy development . . . as indicated by mean scores, and large standard deviations."[76] The photographs could not of course depict the movement of the wind turbines, thus the study contained an "unavoidable drawback." Nevertheless, the panoply of responses unmistakably showed little unanimity in attitude, except that almost all respondents found Altamont wind development "a highly conspicuous, man-made landscape." They were asked whether they liked or disliked the wind generators, whether the turbines were appropriate or inappropriate, boring or interesting, attractive or unattractive, efficient or inefficient. The report said answers were very much determined by respondents' overall impression. Thayer and Freeman concluded that "wind developments of the scale of Altamont are new landscapes with little existing precedent. It remains to be seen how public attitudes toward them will change over time."[77]

Other researchers worked on the crucial area of bird deaths. This was a problem with no easy solution. The red-tailed hawks and golden eagles of Altamont, which soar on updraft winds and find prey on the grass-covered hills, had vocal, determined human supporters. When development commenced in 1980, few people anticipated raptor deaths from collision with wind generators or electrocution. However, between November 1984 and April 1988, ninety-nine such incidents were reported from Altamont Pass. During the same period, researchers documented only nine avian deaths at Tehachapi Pass. The Altamont deaths broke down into thirty-six from electrocution and sixty-three from collision. Observers found electrocuted birds at the base of power poles, indicating contact with uninsulated conductors and wires. In response, the wind-energy companies insulated exposed wires and initiated measures to minimize the chance of such fatalities. A 1989 CEC report indicated that "corrective measures have been successful where implemented."[78] But the collision problem was more intractable. Descriptions of "head, leg and body injuries . . . including completely severed heads and bodies sheared in half" made it clear that the turbine blades did the damage.[79] Remedies such as painting the blades and installing sound-warning devices (actions that might reduce bird deaths) were not acceptable: they would create more visual blight and noise pollution. Some researchers have suggested that changing from lattice towers to tubular, thereby reducing bird perches, would make a difference. On this issue of bird mortality at Altamont, David Nesmith, of the San Francisco Bay Area chapter of the Sierra Club, stated:

> We recognize that there are environmental trade-offs in any form of energy production. Whenever we think we've found the perfect environmentally benign form of energy production we find that there's a problem. We're certainly supportive of . . . [wind] as less destructive than fossil fuels and nuclear energy production, but the raptor damage certainly disturbs us.[80]

For Nesmith and many others, the matter of energy production is not one of positive good but one of necessary evil. The key is to find a sustainable, reasonably benign energy source.

LIVINGSTON, MONTANA

If a frontier is the edge of the unused, California represents the heart of the new wind-energy frontier. The direction of spread of the wind-energy business finds a parallel in history. Twentieth-century wind energy is comparable to the nineteenth-century mining industry: it commenced in California and moved eastward. Although, as we have seen, wind-energy development has roots in the eastern United States and Europe, there is no denying that the great surge of development has occurred in the far West—even as far west as Hawaii.

In the nineteenth century, the gold strikes of California were followed by rich diggings on Grasshopper Creek and the Virginia City area of southern Montana. And in the early 1980s, it was a small Montana town that decided to replicate the California wind experiment. One of the more absorbing stories of wind energy involves Livingston, a community of some five thousand persons, nestled to the north of the Absaroka Mountains, along the banks of the Yellowstone River. The town profited as a gateway to Yellowstone National Park. Besides tourism, ranching provided employment, as did the Burlington Northern Railway diesel-engine repair shop. Community complacency was disrupted in the spring of 1976 when Burlington closed this repair shop, eliminating more than three hundred jobs. Many of the employees transferred to other Burlington facilities in Burlington, Iowa, and Alliance, Nebraska. The closure devastated Livingston. With unemployment depressing both the economy and morale, people began to think anew about the available resources.

Livingston had long been known as one of the breezier places in the nation, but the constant winds were considered to be an annoyance. The Chamber of Commerce did not draw attention to Livingston's winds. However, with jobs scarce, options few, and energy expensive, this unexploited energy source presented possibilities. The nation's growing concern regarding oil dependence gave wind-electric generation an almost patriotic mandate. When Congress passed the PURPA Act (1978), guaranteeing a market for independently-generated electricity, the stage was set for community action.

The person who turned the idea into reality was Ed Stern, who had been hired as a development officer for the town. Stern approached his task with enthusiasm, determination, and political savvy. He soon convinced the city fathers to restore the downtown area and then turned his attention toward energy production. What could be done with the old airfield site? He suggested wind turbines. Getting the town behind such a novel idea was not guaranteed. Stern talked incessantly. Armed with an attractive slide presentation, he solicited clubs, churches, and anyone who would listen: ranchers might profit by leasing their land; the city would benefit through new jobs; the community would be united in an environmental enterprise. His enthusiasm proved infectious. Stern, as Montana State University professor Gordon Brittan put it, had "a vision of a bright new tomorrow."[81]

The idea of a city wind farm was unique. The city airport had recently abandoned its location east of town and the site was available—and thus did Livingston become the first municipality to be directly involved in the wind-energy business, promoting the idea, leasing land, and, hopefully, sharing in the profits. The whole project was civic in character. To carry it off, the city petitioned the Environmental Protection Agency (EPA) and was awarded 85 percent of the total project cost of $355,000. However, in 1981 a new federal administration (Ronald Reagan) froze the EPA grant. The city reduced the scope of the project from eight turbines to four and Stern then appealed to the Montana Department of Natural Resources and Conservation, seeking a grant from the alternative renewable energy sources fund. Livingston was granted full funding for a $222,000 wind farm.[82]

A number of developers expressed interest in the site. When the Montana Power Company negotiated a total buy-back rate of 6.43 cents per kilowatt-hour, Jay Carter Enterprises, a Texas firm working hard to be a leader in the field, took up the challenge. Amid considerable local enthusiasm, company engineers erected five two-bladed, 25kW machines. Nearby, Montana Power put up its own 25kW machine. The experiment was under way. However, the Carter machines soon ran into generator breakdowns and the cracking of fiberglass blades.[83] A final, dramatic event occurred on a cold, blustery winter day. A hydraulic-brake mechanism failed and one of the machines destroyed itself, the blades spiraling over the landscape.[84] Carter Enterprises tried to repair or replace the machines, but when failure followed failure they left the field.

Another corporate interest took up the challenge—an investment group that introduced Enertech machines from Vermont. Capitalizing on investment tax credits, they raised five turbines on a Bureau of Land Management site adjacent to the city land. A large crowd gathered for the start-up. A brisk breeze set the blades turning, but within five minutes all five machines were down, stymied by a simple mistake: the wrong gearbox fluid had resulted in overheating. The Enertech machines never did get on-line. Not only were they victim to human error and insufficient engineering; disinterested investors and the uncharitable Montana winds also played a part in their demise.[85]

The next venture involved an Illinois man, Greg Cook, head of a company called Windpowered Machines, Ltd. Cook had designed a unique sail windmill, which strongly resembled the ancient windmills of the Lasithi Plains on the island of Crete. The Windjammer, as it was named, turned slowly, featured understandable and appropriate technology, and with its Dacron sails was pleasing to the eye—far more so than the popular Danish three-bladed mills. Cook had already experimented in Illinois with a twenty-foot 5kW machine financed by International Harvester; now he was prepared to erect a seventy-foot 105kW machine at Livingston.[86] The idea attracted local investors and in the summer of 1983 Windjammer 2 began generating electricity, feeding the Montana Power grid.

Turbine design is still in its infancy. The Windjammer 2 machine turns slowly and is reminiscent of ancient windmills of Crete; however, the prototype is unreliable. (Courtesy Gordon Brittan, Montana Wind Turbine Co.)

However, Cook faced financial difficulties. One day he disappeared mysteriously, leaving investors to take over the patents and form the Montana Wind Turbine, Inc. Since that time, the company has continued to work on the design, meeting with some success and considerable failure. Company directors still hope to manufacture simple, mechanically workable wind generators, aesthetically unmatched by other designs.

Next to step up at Livingston was Fred Bannister, a dentist from Wisconsin who had a sincere commitment to wind energy. Bannister figured that the new Jacobs 25kW machines, manufactured by Control Data Corporation, could be successful at Livingston, but the new Jacobs turbines proved less reliable than the old Montana "Cadillacs." Bannister, like others in the trade, decided to switch to the more trouble-free Danish design. He formed Wind Power Systems (WPS) and acquired a distributorship for Windmatic, soon persuading a group of Montana investors to purchase two 14-S turbines. However, Bannister was unaware of a crucial factor—the two turbines were untested prototypes. When placed in operation at the city site, they seemed to disprove the touted trustworthiness of Danish products. The whole nacelle of one machine fell off the tower and the other soon shut down with a broken blade. These machines were two of five 14-S prototypes, a model that never went into production.

In spite of the initial failure of the Danish prototypes, interest in them continued. Dr. David Healow, of Billings, retrofitted the two 14-S towers and then installed the more proven and reliable 15-S models. In the meantime, Bannister acquired the inventory of Basin Petroleum, of Wyoming, which had become involved in the manufacture of a 75kW Danish clone and, renaming it the Silver Eagle, installed one next to the Windmatics. Thus, in 1990 five machines harvested the wind—four in the city wind farm area and one Windmatic 15-S on Gordon Brittan's ranch a few miles to the east.[87] These machines have given satisfactory performance for the last few years.

In the meantime, however, the people of Livingston had soured on wind energy. In 1986, following Bannister's initial failure, another outfit (named the Don Quixote Company) had put up a machine to take advantage of the tax credits. It never turned. Ed Stern had made elaborate promises, none of which had come true. Local ranchers had leased land for wind farms that never materialized, leaving them without much-needed income. For a decade, travelers along interstate 90 had seen a variety of machines that never worked. It seemed that the community had suffered from unfulfilled promises, foolishness, fraud, and failure. There were a few winners, but many losers. All the general problems that beset the wind energy business in the early 1980s appeared in microcosm in this Montana town. Perhaps this was exemplified in 1982 when Stern appeared before members of the American Wind Energy Association, extolling the possibilities of a new partnership between local government and the industry but warning industry people to weed out "those in your own ranks who seek nothing more than a fast buck without consideration for quality, ethics or good practices, and not allow them to taint your good name."[88]

Stein has moved on, but the story is far from over. In 1991, Montana Power realized that it would soon lose 150MW of imported power, related to cutbacks announced by the Bonneville Power Administration. The cuts in hydroelectric power were being made in an effort to save diminishing salmon runs. The dwindling of one renewable resource might stimulate another. Montana Power, with the support of the state PUC, put 150MW out to competitive bid. This decision suggested a new climate—the utility company would accept a percentage of independently produced energy rather than build a new power plant and thus maintain a near monopoly on production. The move to solicit cogeneration of power also marked a shift toward decentralization.

In response to Montana Power's call, SeaWest, U.S. Windpower, and the New World Power Company submitted eleven wind-energy proposals for the state. As word spread that hundreds—perhaps thousands—of wind-power units might be lined up across the Yellowstone Bench, the opposition appeared. With reason, many Montanans had turned negative on wind electric and they were supported by newer residents, former Californians, who had seen the visual aberrations of the huge Altamont Pass facilities. The battle was joined. Opponents of wind energy ran a full-page newspaper advertisement proclaiming that wind generators kill eagles "in devastating numbers" and are "dirty, destructive, and irresponsible." Appealing to a Western fear of outside control, the advertisement announced that "wind power companies are one more classic example of an out-of-state power interest raping Montana of its big sky, beautiful rural farmland, wildlife resources, and quality of life."[89] The spred featured a photograph of hundreds of wind generators on the Altamont Hills, warning that "Rural Montana Could Look Like This Soon . . . Unless You Help."[90] Some of the most vocal opponents of the U.S. Windpower proposal were the former Californians, who did not want their new state to be urbanized and the landscape "industrialized."[91]

A public hearing in Livingston in April 1992 amplified the opposing positions. Organized by an anti-wind group named the Park County Environmental Coalition, officials from U.S. Windpower were invited to defend the company against charges that it had destroyed the quality of life in the Livermore area of California and killed eagles to boot. Citizens expressed their outrage. The company might destroy "one of the most beautiful valleys in the state and quite possibly the world." Wind-power advocate and philosophy professor Gordon Brittan assured the hearing that U.S. Windpower would not create another Altamont Pass in the Yellowstone Valley. In reply to criticisms about bird deaths, Brittan minimized the problem. He reminded critics that the Exxon Valdez oil spill killed an estimated 350,000 birds, a figure that made the avian mortality figures at Altamont relatively insignificant. William Whalen, vice president of U.S. Windpower, also tried to placate fears. A past director of the National Park Service, Whalen guaranteed that his company would conduct wildlife studies and aesthetic-value studies in advance of any construction. "If the studies show the machines should not be there, we're out of here," he promised.[92]

Many people remained skeptical. The opponents were unimpressed with Whalen's credentials, noting that before going to National Parks—a political appointment—he had worked as an insurance salesman. Ron Wiggins, a Big Timbers resident and outspoken critic of wind energy, believed that Brittan's use of the

Exxon Valdez figures was ridiculous since the oil spill was the outcome of criminal neglect and accident. The avian deaths at Altamont, and by extension any that would occur on the Yellowstone Plains, were no accident.[93]

Further possible development of wind power in the Yellowstone Valley is on hold. The Montana Power Company rejected all eleven wind-energy bids, largely on the basis that cogeneration from natural gas and waste-to-energy plants was more cost effective. U.S. Windpower has withdrawn temporarily, but remains attentive. Opponents remain vigilant; proponents regret the failure of Livingston to use an abundant, clean resource. In the northern Rockies, land use disputes happen all the time. The residents understand trade-offs between energy needs and a pristine environment. Most would probably sympathize with environmental writer Todd Wilkinson, of Bozeman, Montana who wrote that "harnessing the wind in the right places would relieve the pressure to drill oil in the wrong ones—the Bridger-Teton National Forest and Alaska's Arctic National Wildlife Refuge."[94]

MINNESOTA

Other cities, too, are associated with wind-energy development. Among the more recent developments is that of Minnesota Windpower, Inc., a company founded by Midwest wind-energy pioneer Dan Juhl. Juhl and his partner Lars Olsen (owner of Wind World A/S, of Skagen, Denmark) manufacture the Windharvester/35kW and recently entered into a successful union with the town of Marshall.

The two entrepreneurs approached city officials with the idea of building a small plant of five Wind World 120kW machines that would produce one million kWh annually for sale to Marshall Municipal Utilities, a company serving six thousand customers. The city agreed and signed a ten-year contract to purchase the electricity at five cents per kilowatt-hour. Once officials had signed the contracts and workers had raised the turbines, the town called for a celebration. On May 19, 1992, Governor Arne Carlson and Dan Juhl threw the switch, the turbines coming quickly on-line in a 25mph breeze. The governor praised the project as "a giant step forward in harnessing nature." State Senator Janet Johnson, playing on the title of a Hollywood movie, won the crowd's approval with: "It is not a question of choosing between the environment or jobs. Windfarms are good for both of them. A field of dreams for me is a field of wind machines and if we build them, I believe the jobs will come."[95]

Wind energy in Minnesota ironically was boosted by the troubled nuclear utility industry. In the early 1990s, the Northern States Power Company applied to the state to expand storage of nuclear waste at its Prairie Island facility. The Minnesota legislature granted the request, on condition that the utility company agree to either purchase or generate 425MW of wind-generated electricity. The legislation jump-started the wind industry in Minnesota and in May, 1994, the first 25MW of power went into production. Seventy-three Kenetech 33M-VS turbines are scattered along the horizon of Lake Benton in southwestern Minnesota. Unless the legislature changes its requirement of 425MW, construciton is guaranted well into the next century. Minnesota Windpower's claim that its Marshall project "is the largest commercial windfarm application to date outside the state of California" is presently contestable; however, if construction continues at nearby Lake Benton, Minnesota will become the state with the second biggest wind-energy production in the nation.[96]

HAWAII

In Hawaii—a region very dissimilar to Minnesota—the Hawaiian Electric Company has expressed interest in wind energy since the early 1980s. This makes sense. The islands import large amounts of petroleum, the source of about 90 percent of the state's energy.[97] With strong and constant trade winds, Hawaii had exceptional potential. In the late 1980s, the islands of Oahu and Hawaii, in particular, became utilitarian laboratories. On Oahu's north shore, Hawaiian Electric Renewables Systems (HERS), a wholly-owned subsidiary of Hawaiian Electric Industries, installed a 9MW wind farm, using fifteen Westinghouse 600kW turbines. At Kahuku, January 1988, the Boeing MOD 5B, the last and most advanced machine from the federal wind-energy research program, had been installed at a 12.2MW wind farm and had completed testing. HERS assumed ownership on January 15.[98]

On the island of Hawaii itself, 198 Jacobs wind machines (3.4MW capacity) were installed at Kahua Ranch, and then a bank of 120 Jacobs turbines (2MW capacity) was put up at Lalamilo. The Hawaii Water Department purchased the power. The most impressive installation on Hawaii—the Big Island—was the Kamao's Wind Farm at South Point, where thiry-seven Mitsubishi model MWT-250s were capable of 10MW.[99]

It is useful to ask how wind energy has fared in the Islands. More specifically: How does the only utility company that has itself owned substantial wind-powered generating capacity (12MW) view the experience? The answer is, with guarded optimism. The main problem has been that of reliability, particularly of the Westinghouse 600kWs. In July 1986, one of the turbines lost a blade, necessitating a shutdown of all units for the accident to be investigated.[100] Within a month, the turbines were back on-line, but not for long. Gearbox problems caused intermittent shutdowns until mid-October.[101]

In 1992, HERS officials shut down the fifteen Westinghouse 600KWs as well as the Boeing MOD-5 (3.2MW) turbine. Al Manning, former president of HERS, cited as reasons the chronic mechanical problems as well poor financial performance. Paul Gipe put it more succinctly: "They got lemon turbines."[102] Manning explained to a journalist that the 600kW units "were the only ones of their kind in the world. When parts were needed . . . we frequently had to have them made."[103] Manning listed other reasons for the shutdown decision, among them the 600 kW unit's initial high cost, the difficulty of obtaining a crane for repairs, and the incompatibility of size. "These turbines were created when the industry was still experimenting with turbine size," wrote Manning. "Today's turbines offer a more realistic capacity range."[104] Hawaiian Electric Industries has since sold the MOD-5B and the 600kW turbines to a private company, New World Grid Power, Inc. and in mid-1995, the Boeing was still operational and a few of the Westinghouse 600s were producing. However, it is unlikely that Hawaiian Electric Industries will continue their experiment. The company has soured on the behemoths and if executives further consider wind energy it will be with regard to the new generation of medium-sized, more reliable turbines.

Another problem for HERS has been the Hawaiian Electric Company grid system, which can accept a limited amount of wind power. Particularly on Hawaii, where 15MW of wind power is available to go on-line, the company restricted the maximum to 4MW. Wind-energy company executives were concerned that such a dis-incentive would damage future development plans.[105]

Hawaii presents other unique challenges. An important factor is that of saltwater corrosion. Many of the best sites are on the coastline, catching the trade winds coming off the ocean. The coastline provides good wind free of turbulence. But the problem of severe corrosion has influenced engineers to place turbines inland. However, in a further twist, Robert J. Hutchinson noted that the least corrosive air is not necessarily inland but "is usually found about 100 feet above the seashore."[106] The wind three hundred feet inland is more corrosive. Hence, the optimum placement of wind turbines is right at the shore's edge—or, even better, offshore. However, science and aesthetics often clash. As elsewhere, the placement of wind turbines on the coastline of Hawaii will face stiff opposition.

In some respects, Hawaii is the optimum site: a region of good winds, gentle climate, and virtually no fossil fuels resources. Yet the state's megawatt capacity of wind energy declined in 1992. Zond Systems, Inc. participated in the Molokai Windpower Demonstration Project, a diesel-engine/wind turbine water-pumping project, by installing three Z-16 modified wind turbines. The system worked well, but the utility terminated the project. The future success of wind energy in Hawaii will require greater commitment by the utility company than is now evident.[107]

COMING OF AGE

By the close of the 1980s, wind energy had come of age. For the first time, Americans realized the promise of significant energy production from a benign, nonpolluting, renewable power source that did not create global warming. But it was equally apparent that natural advantages, such as those present in Hawaii, did not necessarily translate into a successful program. Conscientious planning, tax credits, and regulatory incentives were every bit as important. In that respect, California offered the industry a nurturing environment. "Wind power did not choose California as its womb; California choose to foster wind power," reported the *Windpower Monthly* at the close of the 1980s.[108] Inspite of there being many naysayers, there was cause for confidence. Two major earthquakes, one in southern California and one in northern California, had done minimal damage to the turbines. For three days after the October 17, 1989 San Francisco quake, the Altamont wind farms were shut down, but only because the PG&E grid was so disrupted that it could not receive the power.[109] The major Los Angeles earthquake appears to have not damaged wind installations.

A harmonious, indeed aesthetic, image of wind generators is caught by a photographer at Altamont Pass, California. (Courtesy Neill Whitlock Photography, Dallas, Texas)

Perhaps most heartening were the energy-production figures. The California Energy Commission reported that although the installed capacity decreased by 14 percent in 1988, the electrical output increased from 1.7 billion kWh in 1987 to 1.8 billion kWh. Furthermore, the "average capacity factor" had increased significantly: 20 percent at San Gorgonio, 17 percent at Altamont, and 15 percent at Tehachapi. The commission's Sam Rashkin believed that the figure for Altamont would have been significantly higher if the Fayette and FloWind companies had not experienced financial difficulties, resulting in 300MW of turbines producing very little.[110]

The California experiment proved that wind energy could play a significant role in production. At the turn of the decade, turbines were much more reliable, although the public did not always realize that fact.

The most important questions were still unanswered. Wind energy was wonderful for the future, but what of the present? Should NIMBY protests about visual and auditory pollution be judged sufficient reason to negate the global benefits inherent in wind energy? Should the U.S. public be constantly shielded from the realities of energy production? Should we be visually offended rather than globally threatened? Wind energy, the theoretical darling of the environmental movement in the 1970s, drew mixed reviews and passionate responses when theory became the reality of a changed landscape.

CHAPTER 12

A PERSPECTIVE ON THE FUTURE

The answer is blowin' in the wind.

—BOB DYLAN

THE DECADE OF THE 1990S BEGAN WELL FOR RENEWABLE ENERGY. IN APRIL 1990, wind advocates scored a minitriumph when U.S. Windpower erected a 100kW turbine in Washington, D.C., on the Mall, practically in the shadow of the Capitol. Washington observed a wind turbine for the first time. The occasion was Earthtech 90, an environmental technology fair held in connection with the twentieth anniversary of Earth Day.

Not only was the physical presence of wind energy there, but before a gathering of Washington dignitaries and the media, the deputy secretary of energy pushed a remote-control button that activated two hundred wind turbines in California, adding twenty megawatta of renewable energy to the PG&E grid. Through a live television picture of an Altamont hillside, the crowd viewed the flock of spinning turbines.[1] The event seemed propitious, perhaps portending or symbolizing the hope that the 1990s would bring forth a successful decade of growth and acceptance after the tumultuous 1980s.

Randy Swisher, the new executive director of the American Wind Energy Association, had been less dramatic but more succinct at the 1989 annual conference, stating: "We really don't know what

A wind generator turns in the shadow of the Capitol in Washington, D.C., reminding legislators in April 1990 of the benefits of renewable energy. The success of wind power depends not only on the whims of the wind but on the caprice of politics. (Courtesy KENETECH Windpower, Inc.)

the future will bring, but the nineties will surely beat the hell out of the eighties."[2]

Will wind energy grow and prosper, fulfilling the expectations and the hopes of environmentalists as well as its promoters? Will it make a significant contribution to the national energy mix? In 1981, Christopher Flavin, energy expert for the Worldwatch Institute, stated that "the technology for harnessing the wind has come a long way in the last decade, but the progress made so far could be dwarfed by the advances in the next 10 to 15 years."[3] His assessment proved correct. By 1995 great strides were evident, nationwide and worldwide. Can Americans expect another dramatic expansion by 2010? Answers to such questions are elusive. Often the most logical response to a question is to ask questions: Will the public accept wind generators, or will the not-in-my-backyard syndrome prevent development of many good wind energy sites? Will public concern about the care of the earth curtail the use of nuclear and fossil fuels? Will the United States (or foreign) companies successfully manufacture a reasonably priced, reliable, efficient, visually acceptable wind turbine?

As in the past, the reliability factor continues to be a significant bellwether to the health of the industry. Fortunately, the new generation of turbines is more dependable. Furthermore, wind-plant operators have become much more skilled in maintaining the turbines. In 1992, some wind plants reached a 95 percent availability level; that is, the average "down time" was only 5 percent.[4] This is a striking statistic. It translates into the fact that when the wind is blowing these generators are available and do their job. However, the new reliability seems lost on the leadership of utility companies.

Often the drawback is perception, not performance. A 1991 study of the utility industry's perspective on wind energy suggested that "utilities not currently involved with windpower projects believe windpower to be costly and unreliable." These companies, unaware of technological progress, had "little intention of including windpower in their future generation mix."[5] Some chief executive officers responded vehemently: "It is cost prohibitive. It's unreliable." Another replied: "I would not build a wind facility with current technology under any circumstances."[6] These executives drew from the experience of the early 1980s. The sins of the first generation of wind generators are being borne by the second and the third. Randy Swisher states that utility companies are "very risk-averse and it takes a long time for innovation to take hold."[7] It will take a span of time and a dose of education to change the opinion of the utility executive as well as that of the general public.

This reeducation effort is under way. Cycles of attitudes and preferences seem to be turning in favor of this youthful industry. Specifically, many utility company executives are taking a fresh look, becoming more receptive to the gentler, more dispersed renewables such as wind energy.

There are many reasons for this subtle turnabout. The Danish wind turbines certainly have been a factor. Generally, they have proven reliable, a vital factor for any utility company supervisor. The Danish machines are presently the standard, primarily because they work, churning out electricity hour after hour, day after day. In the view of many engineers, however, this dominance is temporary. Robert Lynette and many others are convinced that the lighter, high-tech turbines being designed and tested in the United States and supported by the National Renewable Energy Laboratory (NREL) will emerge victorious. Lynette places his faith in his engineers at Advanced Wind Turbines Incorporated (now a subsidiary of Flowind), who are putting the finishing touches on a new, two-bladed 225-275kW AWT-26 turbine. In mid-1995 two prototypes were in operation, and its proponents promised production models by early 1996. If the AWT-26 is erected at good sites, it will produce reliable energy at five cents per kWh.[8]

In Burkburnett, Texas, Carter Wind Turbine was placing bets on Model 300, a horizontal-axis, downwind, teetering-hub, two-bladed rotor turbine with a maximum-rated output of 400kW. According to CEO Jim Moriarty, the "small turbine and slender tower are less visually obtrusive"; moreover, "the most exciting advantage of the Carter lightweight design is that the small turbine can be scaled up to 1.5 megawatts to optimize the power to cost ratio."[9] Other advantages include a low tower-head weight and a tiltable tower that allows the unit to be serviced on the ground. However, Carter has been sold to a British company. The operation has been moved out of the country and it is doubtful that the scaled-up turbine will go into production.

U.S. Windpower (Kenetech), the leader in the field, also believes it has the secret to reliable, inexpensive wind power with its Model 33M-VS, also designed to generate power at five cents per kWh. According to Kenetech vice president Dale Osborn, the new turbine "has been tested in the lab and in the field. It has worked beyond our wildest imagination."[10] These machines are, as one journalist put it, "a marriage of aerodynamics and microelectronics."[11] They feature thirty-three meter blades constructed of two thousand pounds of fiberglass. Hopefully, they will not be subject to blade failures. The variable-pitch blade system will produce power with a nine mph wind and will shut down at sixty mph, a considerable improvement over earlier units that produced only at wind speeds above fourteen mph and shut down at forty-five mph.[12] Although it is too early to make a firm assessment, thus far the 33M-VS has lived up to Osborn's optimism.

If this new generation of turbines performs as promised, analyst Harry Wasserman will be correct when he asserts that "at current prices, no proposed new nuke project can compete with wind, nor most reactors now operating."[13] While the jousting for price advantage between nuclear and wind may be decided, fossil fuels, particularly natural gas, remain highly competitive. However, one should not forget that the U.S. energy mix requires diversity. There is room for natural gas, wind, and a host of other heat or kinetic-energy

The new Kenetech 33M-VS leads a new generation of medium-size generators that promises electricity at five cents per kilowatt-hour. (Courtesy KENETECH Windpower, Inc.)

sources. Wind energy cannot stand alone, and perhaps the highest prediction of its future share of the energy pie is 20 percent. If Kingsley E. Chatton, of U.S. Windpower, is close in his prediction that his company's new, third-generation wind turbine can produce power at a price of about five cents per kilowatt (about one-half the nationwide retail average cost of electricity), the industry can work toward the immediate goal of 10,000MW of on-line wind power by the year 2000.[14] In some circles, the issue is in doubt. None of the new-design turbines has been spinning for long. Paul Gipe is fond of noting that designing and predicting are much different than delivering, and thus far neither the U.S. government research program nor U.S. private enterprise has been able to deliver a significant volume of electric power at five cents per kilowatt hour, although the Kenetech 33M-VS is approaching that goal.[15] Wind-turbine engineers David Eggleston and Forrest (Woody) Stoddard reminds us that rotor aerodynamics is "clearly not for the faint-hearted."[16] A low per-kilowatt price is of little value if efficiency is at the expense of reliability. These new high-tech, third-generation products must be—like the Danish turbines—tough, able to run year after year in competition with traditional forms of energy production. They must win back the confidence of the public. Some in the industry believe that the turbine race has already been run, and the Danish won it. Curt Maloy, president of the Desert Wind Energy Association, believes that the Danish turbines have become the standard and are likely to remain so.

Paul Gipe is, perhaps, the severest critic of the high-tech engineers and the costly 1980s endeavor to build a teetered, two-bladed, megawatt turbine, describing it as "one of the great tragedies. . .in American engineering." Almost every one of the machines "blew up in the field."[17] Gipe notes that many of the engineers were specialists in missiles, giving them expertise in machinery with a very brief life span, hence their emphasis on performance and efficiency but not reliability. "Our windmills have to last," emphasizes Gipe, "competing with other forms of energy production."[18] He gives NREL (previously the Solar Energy Research Institute, or SERI) high marks for its work on air foils, but believes that it should stop spending tax dollars trying to perfect an advanced turbine.

However, nothing is simple. There are always two sides, two opinions. Those who support the development of light, flexible, two-bladed turbines underscore their great economic advantage.[19] Two-bladed machines reduce the cost from 15 to 30 percent over the three-bladed, a very significant factor. Thus will the proponents of two-bladed and three-bladed argue the merits of their respective units. One windpower wit stated that the conflict between the two was analogous to choosing a religion: you don't know which one will get you to heaven, and by the time you do, it's too late.

Conceivably it is time for NREL to consider a new direction, away from engineering and technology and toward the practical environmental, financial, and political issues that the renewable-energy industry must face if it is to expand—indeed, survive. Studies must be undertaken to minimize avian mortality. Even though bird kills cannot altogether be eliminated, government agencies must assist industry representatives to compile the best information. The public demands serious study of this problem, and is entitled to it. Expansion may hinge on the industry's ability to resolve this predicament. A recent report from the Altamont region was criticized by a U.S. Fish and Wildlife Service expert, who determined that the "questionable study" asked for a "negative declaration" based "on non-existent mitigation measures."[20] Avian mortality has the potential to stop the blades, and alienate the very people who are friends of the industry. Perhaps the most promising avenue of exploration is the notion that lattice towers invite birds to perch and therefore die, whereas tubular towers may resolve the problem. As yet, the studies are incomplete.[21] The issue must be addressed in a serious, scientific, dedicated fashion. The federal government could assist in this effort.

Another issue that continues to fester is the visual effect of wind farms. Opinions on the topic vary, and more research, such as that

FloWind vertical-axis turbines can be aesthetically pleasing. When the author took this photo, these units were turning in unison, as though performing a kind of ballet with the wind.

provided by landscape architect Robert Thayer, could be sponsored by NREL. American ambivalence to the "machine in the garden" will never be resolved, but research may uncover possibilities. Which style of wind generator is least offensive in particular sites? Montana Wind Turbine, Incorporated, has experimented with a 100kW Cretan-sail style that turns slowly and pleasingly, but the design has not yet been perfected.[22] Vertical-axis machines, such as those developed by Sandia Laboratories and installed by Flowind, can be stunning additions to the landscape. When turning in unison they often resemble mechanical dancers performing a graceful, whirling ballet. Henry Robb, of Sandia, believes the jury is still out in deciding between the vertical (VAWT) and the more popular horizontal (HAWT) design, and that vertical-axis turbines may be advantageous in certain situations and landscapes. To the eye of many Americans, vertical turbines are less disturbing and more tasteful in the rural landscape.[23] Wind turbines will never be fully accepted by everyone, but money spent on research into the issue would be well spent.

Research on siting is also needed. Meteorologists have determined many of the best sites, but what kind of windmill configuration is suited to the site? Most wind-energy developers understand that the 6,500 wind turbines at Altamont will not be replicated, unless new projects are situated in the most uninhabited regions of the nation. In more populated regions (where the energy is needed) the landscape—the open space—must be preserved. Will the Danish model, featuring a scattering of turbines, be the answer? Should common rural ownership of one or more turbines be encouraged to promote community and shared interests? Will small clusters of, say, 1MW to 3MW of wind turbines be most attractive for certain situations and sites? Will vertical-axis turbines be more visually compatible in certain terrains? Finally, must engineers occasionally sacrifice the perfect site for a less efficient one on the basis of aesthetics and public concern? Perhaps. Perhaps. Present and future research must give wind-energy developers data—or at least discussion— on such questions; otherwise, all the technological progress of the decade will result in mere paper plans.

The expansion of wind turbines in the next century will depend on the effort of government and industry to mitigate environmental objections and the NIMBY phenomenon. There is evidence, however, to indicate that the U.S. perception of wind energy may be changing. Robert Thayer tentatively believes that a U.S. metamorphosis is underway. "When I originally studied wind energy as a landscape/perception phenomenon, I wanted to see if wind farms' intrusion on the visual landscape would be countered by an accrual of positive environmental symbolism," he told me. "After several years of research, I believe the tentative answer is 'yes.' "[24] The evidence may be found in popular culture outlets, such as Hollywood films and television advertising. Rotating wind turbines have emerged as popular icons—symbols of progress, modernism, reliability, and environmentalism. They are often juxtaposed with quality automobiles and reliable airlines.[25]

Changing the public view has—belatedly—become an industry objective. Today the traveler can pick up the "Tehachapi Scenic Wind Farm Tour" brochure, a self-guided introduction, describing the different turbine styles and providing photographic hints on how to capture the spinning turbines on film, admittedly a "challenging [task] even for professionals."[26] Pacific Gas and Electric Company provides a similar guided-tour flyer for the Altamont Pass region. Also, as one drives on Interstate 5, through the turbines, the traveler is invited to tune AM radio to 530 to hear California Governor Pete Wilson explain the contribution of this renewable, nonpolluting form of energy production.

At San Gorgonio, the wind farms are an acknowledged tourist attraction, and freeway travelers with cameras often pull over for a few photographs and to gaze at the surrealistic scene. Presently, the Desert Wind Energy Association is encouraging the state to construct a turnout to safely accommodate the curious. Children find a fascination with these natural-energy converters. One Minneapolis-based newspaper columnist notes that "when our cross-country travels take us to grandma's house in the San Joaquin Valley of California, one of the key attractions along our route from the airport is a wind farm on a mountain pass east of the Bay Area," and she continues: "My kids challenge each other to be the first to sight these high-tech wind generators of the 90s."[27]

People's attitudes do change. Few would suggest that wind energy should be the dominant energy-production method or that rural America should be covered with turbines. However, those who find the turbines, whether American or European, unattractive,

Often lost in the debate over wind generators is the fact that they provide jobs—although some of the projected employment gains were lost when Danish machines came to dominate. (Courtesy Thomas Braise, WindMaster, Inc.)

may often admit that they are an acceptable price to pay for a sustainable energy source that may add to human longevity on the planet. A wise rural philosopher wrote in the 1980s that "energy is not just fuel. It is a powerful social and cultural influence. The kind and quality of the energy we use determine the kind and quality of the life we live."[28]

Again, my theme is that the future of the wind-energy field is a matter not only of engineering but of the social sciences and the humanities. Many fields of knowledge must make contributions if barriers are to be overcome. The tendency of the engineering community is to knock them down, but it is time to consult those who would quietly dismantle the barriers brick by brick, bringing to bear their expertise in other disciplines and endeavors. Experts should be consulted and their research should be supported. U.S. Windpower contends that its new 33M-VS turbine "could expand U.S. wind electric potential by as much as ten times."[29] Perhaps the technology will be ready, but will there be available sites? There may be windy sites devoid of people and bird life, but that is not necessarily the solution. The industry should not be relegated by default to the most desolate places in the nation, far from transmission lines and consumers. Instead it must rethink how people perceive landscape and technology within that landscape. How can Americans come to accept—dare we say appreciate—wind turbines and the fact that, for a little visual alteration, they contribute to our lifestyle and our longevity on this planet in ways that the massive nuclear, oil, gas, and coal plants cannot match. As earlier noted, understanding and then changing perceptions of this will take great skill. Research must support and industry must honor individuals and disciplines that attend to the considerations of cultural history, cultural anthropology, rural sociology, and, above all, landscape architecture.

Economic considerations cannot be ignored, of course. Whatever the environmental advantages of the soft energy path, in the foreseeable future, wind energy must compete in a world dominated by traditional measurements of energy cost. First, there are fixed charges—charges that are not related to the amount of energy produced. These would include the amortized cost of the production plant, its maintenance, insurance, taxes, stockholder dividends, and finally decommissioning. Second are the operating costs, to include salaries for plant personnel, supplies and materials, and outside consulting fees. Third are the fuel costs, which include not only the cost of fuel but the expense of dealing with the waste and pollution that the fuel generates as an unwanted byproduct.[30] It is in the third category that solar energy shines.

As constantly promoted in the 1930s, the wind is free (as is the sun)—the factor that has been an attraction for parsimonious Americans for many years. However, the advantage of renewables, such as the wind, is not only in initial fuel costs but in waste-disposal costs. Total fuel costs must include the disposal of the spent fuel. Nuclear-energy proponents have long faced the extraordinary cost of radioactive waste, an unwanted byproduct that has brought nuclear-energy development to a halt. Wind turbines create no waste problems. Although they extract energy from the wind, turbines leave the forces of the environment unchanged. However, it is not totally benign. Avian deaths, visual pollution, and noise considerations may, taken broadly, be considered as waste problems. Also, worn-out turbines must be taken out of service. Nevertheless, when the waste considerations of wind energy are measured against those of nuclear or coal-fired plants, there is no comparison.[31]

The weak spot of wind-energy production is in operating costs. The difficulties of wind-energy technology have created a reputation—often well deserved—for unreliability. Since the 1920s, manufacturers of wind-electric systems have struggled against a reputation of building shoddy machines that break down. Furthermore, manufacturers have overrated the power output of their turbines; thus, inferior engineering and false claims account for a measure of the historic reluctance of industry or the public to embrace wind energy. But the inherent difficulty of capturing a capricious energy source must also be recognized. The wind, always changing in force and direction, requires a refined technology that can adjust to quick and constant change. The power in the wind increases by the square of the wind speed. If wind speed doubles, wind power grows eightfold; when it triples, wind power is multiplied twenty-sevenfold.

Capturing this force is a great technological challenge, but one that research and development is meeting. Edgar DeMeo, in charge of renewable-energy research at the Electric Power Research Institute, recently stated that wind turbines are "beyond Kitty Hawk and into the jet age." DeMeo characterized the so-called pioneer machines of the early 1980s as crude and flimsy, and largely no longer in use.

Using the new technology, a number of utility companies and energy developers have proposed projects and some are under way.

In a developing field, to list potential projects is to invite being out of date even before the ink is dry on the paper. For instance, in the early 1990s Kenetech Windpower announced a project with Iowa-Illinois Gas and Electric Company to build wind-electric generating stations equaling 250MW in the northern Plains states of North and South Dakota and Nebraska. However, the utility company cooled to the project, and at present the idea is dead. More hopeful is another proposed Kenetech project in southeast Wyoming, near the Medicine Bow site, where Kenetech Windpower proposes to erect 1,390 33M-VS turbines over a ten to twelve year period—a 500MW total.[32] Perhaps more firm is the decision of the Sacramento Municipal Utility District (SMUD) to install 50MW to replace the troublesome (now decommissioned) Rancho Seco nuclear plant. These wind generators were to be built by Kenetech on the Montezuma Hills of Solano County. SMUD will own the 3,400-acre site and the 184 units. By committing itself to ownership, SMUD has resolved the major barrier of financing. Furthermore, in the mind of environmental writer Peter Asmus, the SMUD proposal "is significant because of its unprecedented reliance on local public opinion to choose new power plants. Ratepayer balloting, and numerous public hearings indicated strong support for a diverse power supply system that included renewables such as wind."[33] An enlightened company, which at one time was led by S. David Freeman, a key architect of President Jimmy Carter's national energy strategy, is moving toward reliance on renewables with the approval of its customers.

In the Pacific Northwest, the Puget Sound Power and Light Company has supported Robert Lynette's efforts to build an efficient, reliable turbine, and has discussed a facility of 275 turbines (28MW). The Bonneville Power Administration, in a precedent-setting move, took bids for four possible wind-farm sites in the Northwest.[34]

Interest is perhaps strongest on the Great Plains, which in an earlier age was a vast haven for two million water-pumping windmills. In spite of the doubts and mechanical failures at Livingston, the potential for Montana is immense and projects are being proposed. To Montana's east, Randy Swisher, director of AWEA, reported that North Dakota has such potential that it may be labeled "the Saudi Arabia of wind energy."[35] On the south wind velocities drop off, but not by much. Potential production in such states as South Dakota, Wyoming, Nebraska, and Kansas is impressive, ranging from 27 percent to 40 percent of each state's energy needs. Perhaps the wind in West Texas offers the most stunning future potential. Wind-energy leaders such as Vaughn Nelson, at West Texas State University, and R. Nolan Clark, of the U.S. Department of Agriculture, have long recognized that the Panhandle region could change the method of Texas energy production. Some 44 percent of the state's energy requirements could be met through extensive development.[36] Andrew Swift, a professor of engineering at the University of Texas at El Paso, is even more optimistic. "Just by utilizing the state's ideal wind resource areas like the Davis Mountains and the Panhandle," predicts Swift, "we could generate

The new generation of Jacobs units can still be seen at work. This unit provides power for a rest stop on Interstate 25 in New Mexico. (Author photo)

more than the total electric energy needed by the state."[37] Some companies are seriously looking at this potential. Windsmiths may soon be replacing oil's roughnecks in the vast reaches of west Texas. In the Davis Mountains, north of the Big Bend country, Central and South West Corp., of Dallas, is planning to erect twenty turbines (6MW of power potential). These turbines will be purchased by the utility company, making it the largest utility owner of turbines in the country.[38] But the Davis Mountains project will be dwarfed by nearby Delaware Mountains project, a 250MW project to be developed by Kenetech. If these projects come on-line, Texas will quickly become a major producer, second only to California.[39]

All across the nation, there are hopeful signs that a wind-energy renaissance may be in the offing. However, caution is suggested: the financial realities, the existence of a reluctant utility industry, and the market question (examined at the end of this chapter) weigh against an optimistic outlook.

THE POTENTIAL FOR SMALL UNITS

Manufacturers of independent, small-scale wind-energy conversion systems (SWECS) would be well advised to proceed watchfully. Small units have played a less significant role in the last decade and only a few manufacturers survived the decline and loss of tax credits in the mid-1980s. The U.S. market is limited. Companies such as Wind Turbines Industries Corp., which manufactures a new version of the old Jacobs in five models ranging (from 10kW to 20kW) have a limited market in remote areas. Manager Steve Turek places his faith in the Class A Hybrid System, which combines a Jacobs-type wind turbine with an array of solar panels and a lead-free stationary battery.[40]

Perhaps the most successful manufacturer of small units is Bergey Windpower, a family-run firm. Karl is chairman and his son, Michael, is president. Since its founding in 1977, the company has produced more than 1,200 units, and Bergey 1kW and 1.5kW units may be found in practically every state. In the last few years a larger (10kW) model, the BWC Excel, sold well. Perhaps Bergey's most notable effort has been overseas. Bergey has sold units in about fifty foreign countries, installing hybrid remote systems in towns at the foot of the Himalayas and in a rural commune in Morocco.

Paul Gipe works on a Bergey wind generator at Tehachapi Pass, California. Producers of small scale wind generators hope to sell units to Third World nations. (Courtesy Paul Gipe and Assoc.)

Innovative ideas include the 1.5kW unit, which is designed for easy upkeep. Village Level Operation and Maintenance (VLOM), a standard established by the World Bank, assures buyers in a developing rural community that, with training, local users can maintain and repair the system. Bergey also encourages overseas manufacturing of its units, which provides local employment and lower production costs. Production has been authorized in India and Australia and the company is considering similar arrangements in other countries.[41] Bergey is on the right track. If financing is available and reliability is assured, the foreign market is limitless.

Mexico offers an important opportunity for manufacturers. Recently, Vaughn Nelson, a physicist who heads the Alternative Energy Institute at West Texas State University, visited Mexico and reported that the country has 15,000 nonelectrified villages, a potential market of $50 to $100 million in rural electrification. Wind systems could be competitive in at least 25 to 50 percent of the sites.[42]

EUROPE GETS ITS SECOND WIND

Worldwide, it is less necessary to speculate on what might happen, because in Europe it is already happening. European wind-energy capacity has increased dramatically. As mentioned, in the mid-1980s the California wind farms produced approximately 90 percent of world wind-power generation. However, since 1990 Denmark, Germany, the Netherlands, and the United Kingdom have expanded their efforts. According to one account, by the close of 1992 the U.S. share of wind-power generation worldwide had been reduced to 67 percent. If the European Community commitment to a capacity of 8,000MW by the year 2005 is anywhere near close, the U.S. share will be reduced further.[43]

Truly, the Old World is experiencing a second wind. A 1992 wind-technology status report by the American Wind Energy Association states that "European development is now expected to dwarf that in the United States throughout the remainder of the 1990s."[44] How far the trend to renewables will go no longer hinges on technical issues so much as cultural acceptance. In the Netherlands, a land in love with its windmill tradition, the wind turbines have sparked vigorous debate. The Dutch do not have the same feeling for the new wind turbines that they hold for the cultural icons of the past. It is an important issue. A University of Amsterdam professor has noted: "The future of wind energy will hinge on the public accepting the visual impact of wind turbines as the small price to pay for clean energy."[45]

THE MARKET QUESTION

The future development of American wind energy has taken on an ironic twist. With the great riddle of reliability close to resolution, the wind-energy producers face the prospect of having no market for their product. In California, the ten-year SO4 contracts have expired. In such an unsettled situation, in 1993 the Canadian Province of Alberta installed more megawatts of wind energy than did the whole of the United States.

Survival, not growth, is the hope for some California producers. In a depressing analysis, the American Wind Energy Association's 1992 Wind Technology Status Report admitted: "California wind developers have reached the end of the line for now. They have exhausted all remaining unused utility contracts." The report continued, "Currently, there is only one firm contract for new wind generating capacity in the U.S. (and that for only 5MW) from 1992 through the year 2000."[46] Zond Systems has shut down machines, not because they are broken but because they lose money. San Gorgonio, Tehachapi, and Altamont wind farms all have turbines that are turned off, put out of work by the low avoided cost rates paid by the utilities under the new contracts.[47]

Throughout the 1980s, California encouraged the growth of wind turbines by what the free enterprise system would call artificial means. Tax credits, a sympathetic state government, and a guaranteed market nursed an infant industry. By the late 1980s, however, the helping hands were removed, and it now appears this energy offspring can stand on its own feet. But will it be allowed to? PG&E and SCE have been playing hardball over one, two, or three cents per kilowatt-hour. This debate over avoided cost—so potentially destructive to the industry—will seem insignificant in the future. Birger Madsen, a Danish wind-energy consultant, caught the folly of it when he commented: "I would think that by the middle of the next century people would smile at the discussions which still raged in the 1980s over renewable energy and whether it was 20 per cent or 50 per cent more expensive than fossil fuel generated power."[48]

One can argue that the current crisis is merely a working out of American capitalism—a survival of the fittest. Those companies that are well funded and have little debt will survive, the weak ones will not. There is merit in such a system: plenty of inefficient, often unscrupulous, companies have been weeded out in the past decade. However, there is a time when the merits of Darwinian economics are outweighed by its evils. Energy production has rarely operated on pure capitalism. Investor-owned utility companies represent regulated monopolies. The great utility companies have coal-fired plants that are inefficient, but until they can be phased out, their inefficiency is absorbed into the rate structure. Wind-energy companies should not be held to a standard that the utility companies themselves cannot meet.

COUNTING PENNIES

It is apparent that PURPA established a new industry but did not insure its success. Put another way, PURPA guaranteed a market but not a price. It is the state public utility commission that establishes the price—the avoided cost. Through the power of pricing, a public utility commission can encourage renewable energy development or it can thwart it. Some commissions, such as California's, have stimulated the growth of "qualifying facilities" by establishing a reasonable avoided cost rate, averaging about 7 cents per kWh. Some commissions insure against the development of renewable energy by establishing a low rate. In Colorado, the public utility commission has set the avoided cost rate at 1.9 cents per kilowatt-hour.[49] Such a rate is a major dis-incentive to wind-energy development. Even with the 1.5 cents per kilowatt-hour federal subsidy provided under the 1992 Energy Act, development in Colorado would rate as fiscally foolish.

Rather than fight for higher rates, the industry has tended to accept rates determined largely by the politically volatile and environmentally nonsensical price of oil. Why should the price of petroleum or natural gas determine the future of renewable energy? What is the logic or wisdom that suggests that the life or death of wind energy should be tied to the vagaries of global or domestic markets that are dominated by diplomatic posturing, profit, and the short view? Yet that is the system of today. It is a rationale of fear and it is used effectively by the utility companies, who—as noted earlier—constantly remind California wind-energy producers that since the early 1980s, when most SO4 contracts were signed, the price of oil and natural gas has dropped. Now, according to Robert Scheffler of Southern California Edison, the CPUC tends to favor renewal of the contracts as long as that renewal is not at the expense of the ratepayers, i.e., the public.[50]

Protection for the public is the traditional role of a public utility commission. However, if protection of the ratepayer results in the devastation of the wind-power industry, advocates of that industry have to question if such a result is true protection. One is reminded of the adage, Pay me now, or pay me later. For the saving of a few cents today, the consumer may pay a high price in health, a liveable environment, and added expense tomorrow.

The pricing issue is crucial. The industry, the sympathetic public, and federal agencies such as the Federal Energy Regulatory Commission (FERC) must fight for a more reasonable, higher avoided cost rate. As the contracts expire and the eleventh-year issue takes hold, the new floating rate will make or break companies. "We could be wiped out overnight by the stroke of the pen," warns Paul Gipe.[51]

Into early 1995, the industry had been ineffective in the few efforts made to break the narrow, myopic view of the avoided cost. Instead of arguing on the basis of environmental concerns and the needs of future generations, many have played "the penny game." A company will predict that its new turbine will produce energy at five cents per kilowatt-hour. The wisdom of predicting so few cents per kWh for phantom projects is questionable, since it tends to build expectations within both the public sector and the utility companies that cannot be met. Ironically, perhaps the worst offenders of this pie-in-the-sky approach have been officials of the DOE/NASA wind-energy program, who have often predicted wind energy at two or three cents per kWh, but never come close. Such predictions belong in history's scrap heap along with those made by nuclear experts forty years ago—predictions of energy so cheap it would not have to be metered.

Unrealistic predictions of economy and reliability are what ruined the reputation of the industry in the early 1980s. There is no reason for this trend to continue into the second half of the 1990s. Industry leaders, with the support of federal officials, must convince state public utility commissions to adopt a different tune. They should argue the environmental merits of wind, not quibble over pennies. They must contend for the future, stressing that the welfare of the ratepayer must not be narrowly construed. If industry leaders can convince a state PUC to consider the environmental, social, and political costs of the hard energy path, renewable energy will indeed be competitive at ten cents per kWh. Wind-energy consultants Gerald Braun and Don Smith recently argued that although wind technology has the least-damaging environmental effects, fossil-fuel technologies "have significant unrecognized costs, such as health and safety, research and development support, tax incentives, and military support (e.g., to maintain US dominance of oil-producing areas)."[52] Perhaps it is extreme to add the cost of Desert Storm to the kWh cost of oil-fired power plants, but there is equity in figuring the total cost of energy production. These "external costs"—those not paid directly by the users of the energy, but rather assumed by the society at large—are practically nonexistent for wind. A European study places the external cost of fossil fuels at two to six cents per kWh, and six to thirteen cents per kWh for nuclear.[53] If U.S. utility companies and public utility commissions included the social costs (or savings) in establishing the avoided cost rates, these rates would more closely approximate the true cost of energy. Such action would establish a more level playing field and renewables would make substantial inroads into traditional energy production methods.[54]

Wind turbines enjoy other advantages that are seldom noted. In the arid West, wind-energy advocates must calculate the amount of water saved, a significant advantage over other production methods. In January 1993, Michael Marvin, of the American Wind Energy Association, offered not only the water argument to the Texas Public Utilities Commission but also that, with wind energy, there is no fluctuation in the price of fuel. This assists in stabilizing rates. Furthermore, the greater the diversity in technology, the more reliable the system. Regarding the constant concern over power-plant construction, Marvin underscored the advantage of "modularity"—that is, in comparison with traditional coal or natural gas plants, the unit size is small and therefore lead time for planning and construction is short. Construction can be matched closely to load growth, thus minimizing the chance for a mismatch between supply and demand.[55] If these advantages are skillfully presented, public utilities commissioners may begin to remove the hidden subsidies that have historically encouraged the utility industry to build huge, capital-intensive fossil-fuel plants.

Whereas U.S. utility companies and regulatory agencies have been slow to recognize the economic advantages and reluctant to address the concept of the social costs of energy production, Europeans have had less difficulty. That is why, as noted earlier in this chapter, the Europeans are moving ahead rapidly with wind-energy projects. We will do well to remember Birger Madsen's prediction that people in 2050 will smile over the contemporary debate over the cost of fossil fuel versus renewable energy.[56] The false economy will then be seen as through transparency. It will not be far into the future that society, both European and American, will choose the course of environmental stewardship, if necessary willingly paying a premium for energy that can be produced without damaging the environment and undermining future survival.

In the meantime, the wind turbines continue to spin. In 1994, 15,900 medium-scale machines were generating nearly 3.4 billion kilowatt-hours of electricity nationwide. These turbines created enough energy to provide for the year-round residential needs of more than one million people. Although located primarily in California, small projects operated in Oregon, Minnesota, Montana, New York, Wyoming, Texas, Hawaii, and New England.[57] No matter what the future might hold, wind energy had survived the throes of infancy. Perhaps it will contribute significantly toward meeting the energy needs and environmental demands of the approaching twenty-first century.

EPILOGUE

BOTH THE FEDERAL GOVERNMENT'S EFFORTS IN WIND-ENERGY DEVELOPMENT and California's enterprises tempt the historian to place blame for failure and give praise for success. The Department of Energy's (DOE's) failed big-turbine policy, aimed at the economy of scale, is particularly vulnerable, since approximately one-half billion dollars have been spent in the effort between the years 1974 and 1992.[1] However, research and development monies are often used to fund experiments that are too risky for private industry to undertake. Sure bets do not require federal funding. Although the experiment misfired, engineers gained knowledge, particularly about what does not work. It must also be noted that in the politics of subsidy and grant monies, Congress did not bestow excessive funds on wind energy. Like any boom, the wind-energy business—in both the public and private sectors—was replete with the interplay of good intentions and flawed public policies.

Flawed policies also characterized the California wind industry in the early 1980s. Could the disreputable financial schemes and shoddy machinery have been avoided by wiser policies? The answer, of course, is yes. State and federal subsidies could have been tied to the actual *production* of wind-generated electricity. But then we must ask: would investors have financed the wind farms on the condition that they must produce energy to ensure a return on their capital? Financiers would have been more cautious. Many would not have advanced a cent for such a risky adventure. Like the utility companies, large investors are averse to risk: they may not demand a profit, but they do require assurances against loss. Wind-energy development in California would have been sluggish and prolonged, or might not have occurred at all.

Whatever the mistakes of the past, the first phase of the wind energy story is at an end. It is a story of failure and of success, but primarily of rejection. Nevertheless, intermittent advancement has occurred through individual enterprise, corporate interest, and governmental incentives. It is a story of periodic advancement, including cycles of growth and acceptance.

In the pre-electric period, societies depended heavily on wind, water, wood, and coal as primary power sources. However, engineers cast aside these power-producing methods (with the exception of coal), believing them to be inadequate for an industrial nation. In the United States, as the nation neared the twentieth century, coal and petroleum became the primary fuels of choice. These primary sources offered convenience, availability, control, and the economy of scale. The fact that petroleum and coal were nonrenewable and polluting to the atmosphere were not important considerations.

In general, Americans embraced a "myth of superabundance"; that is, there was the feeling that the nation drank from a never-ending fountain of energy. A resource-lavish continent perpetuated this sense of inexhaustibility, submerging any consideration of a concept of finite resources. A few scientists in Britain and the United States considered Lord Kelvin's plea in the early 1880s for wind-energy development to be inviting, but most found it quaint, incompatible with a dynamic, industrial world. Even Charles Brush did not bother to patent his wind dynamo: he understood that such inventions would be swept aside by centralism, complex systems, and the economies of scale.

Following World War I, proponents of wind energy found a market in rural America: thousands of wind turbines provided electricity for the nation's ranches and farms. However, many people considered this intermediate technology to be a passing fancy, destined to be replaced by centralized power systems. Thus, although in the interwar years wind energy enjoyed a brief hiatus, in the postwar years the bubble burst. By the 1960s, the rural wind turbine had joined the ranch's pile of discarded or worn out technologies.

During World War II, a brilliant engineer named Palmer Putnam turned the wind-energy business in a new direction. In Vermont's Green Mountains he raised the first modern turbine, the 1.5MW Smith-Putnam. It performed much better than most first-stage machines, but after two breakdowns engineers abandoned the experiment. What killed the turbine atop Grandpa's Knob? It was not faulty technology so much as a faulty vision of the nation's energy future. Why research a diffuse, difficult energy source when petroleum was plentiful and nuclear power would provide limitless, inexpensive energy in the future? Thus, the visionary ideas of engineers such as Palmer Putnam (and Percy Thomas) were blithely cast out by politicians and scientists who placed inordinate faith in nuclear power, another version of the myth of energy superabundance. During these decades, Congress fed billions of dollars into nuclear research; not a penny of appropriation found its way to the development of wind energy or any other alternative energy source.

In the early 1970s, three factors combined to compel Americans to rethink electrical-energy production methods. Advocates of nuclear energy faced exorbitant costs and concerns regarding safety and storage of waste. More stringent clear-air laws made coal-fired plants less desirable. Most important, the uncertainty of international petroleum supplies, revealed by the OPEC oil embargo in 1973, caused Americans to realize their vulnerability. The rules of the energy game changed. Federal funds flowed into the newly created Solar Energy Research Institute (SERI), and wind-energy plans emerged from the drawing boards of both public agencies and private companies. The DOE embarked on a major effort to develop a commercially viable, large wind turbine, focusing on the elusive economy-of-scale concept. It was unsuccessful.

In California, however, a combination of federal and state incentives, aided by a guaranteed market for wind-generated electricity spawned a burst of activity from 1981 into the early 1990s, resulting in 16,000 wind generators. When Paul Gipe dedicated his wind-energy book, *Wind Energy Comes of Age*, to the "men and women who breathed life into a dormant, though never dead, technology," he meant to honor the many idealists who transformed an idea into a reality.[2]

To return to the present: the wind farms in California are functioning, the result of more refined technology and sounder economic practices. They represent the first significant alternative-energy generating stations in the United States. Furthermore, the lessons learned in California are being incorporated elsewhere. Across the nation, utility companies are designing or contracting for new wind-energy projects. The wind-energy industry is poised for growth.

With the approaching twenty-first century, the second phase of wind energy will begin. Whether it will prosper depends on politics, policies, and pricing. At present the cheapest new method of producing electricity is through combined-cycle natural gas plants. Such plants are relatively clean-burning, but they add to global warming and they use water, a consideration for arid regions. Furthermore, natural gas is nonrenewable. An abundant, cheap fuel, it may quickly become scarce and costly. National myths of superabundance ought not to be forgotten in a rush for the cheapest primary energy source.

The health of the wind energy today is attached to the thermometer of deregulation of the utility industry. The example of deregulation of the airlines and the long-distance telephone industry leads some economists to argue for the same transition to free enterprise for the utility industry. They contend that consumer electricity prices must reflect the most efficient means of energy production, which translates into a swing toward the use of natural gas. Given present fuel prices, producers of energy can build natural-gas plants and sell electric power at 4 to 5 cents per kilowatt-hour, underselling conventional power and wind power prices of 7 to 10 cents per kilowatt-hour. In a national mood of deregulation, it is possible that small natural-gas generating stations can drive the large utility companies to distraction, especially if they are able to distribute—or "wheel"—their cheaper electricity on utility company lines.[3] Wind-energy companies may be caught in this dilemma because many of their contracts are now being negotiated to lower "avoided cost" rates. They could, and some are, being driven from the field.

It is not within the purpose of this book to examine the arguments for and against deregulation. The utility question is complex. Since the days of President Theodore Roosevelt, Americans have embraced public control over monopolies such as utility companies, primarily for pricing protection. However, in recent years many state public utilities commissions have enlarged their responsibilities to include environmental considerations.[4] Although market forces are, in themselves, amoral, the result of leaving events to their direction is that activity flows to the cheapest, most profitable means of energy production, without consideration for the environment or the future. Total deregulation would ignore the limits of our finite world as well as the rights of future generations. It would abandon any concept of stewardship, leaving the fate of the world to chance rather than sensible planning.

A new century and a new chapter in wind energy beckons. Regulators, politicians, scientists, and engineers will grapple with finding the proper combination of primary energy sources: a combination that will be sustainable, reasonable in cost, and the least offensive to both the environment and human perceptions. Decisionmakers must acknowledge the bottom line, but they must not bow to economic considerations alone. The United States is, after all, a nation capable of instituting policies based on not only price but sustainability and a livable environment. A sensible electric-energy production policy will promote diversity—a mingling of kinetic and heat sources and a recognition that some pre-industrial energy sources have a place in the postindustrial world.

NOTES

CHAPTER 1. HUMAN USE OF WIND ENERGY

1. Second verse of Caroline Tappan's poem, "Windmill," *The Dial* 2 (October 1841): 230.

2. M. R. Gustavson, "Limits to Wind Power Utilization," *Science* 204 (April 6, 1979): 13–17.

3. Geologist David Love, as interpreted by John McPhee, "Annals of the Former World," *New Yorker*, (February 24, 1986): 67.

4. Jamake Highwater, *Ritual of the Wind* (New York: Alfred Van Der Marck, 1984): 29.

5. Elsdon Best, *Maori Religion and Mythology*, part 2 (Wellington, New Zealand: P. O. Hasselberg, first published in 1982): pp. 38, 411–3.

6. From E. A. Speiser's translation of the Bible (Genesis 1:1).

7. 1st Kings 19:11–12 (RSV).

8. See Keith Crim, ed., *The Interpreter's Dictionary of the Bible* (Nashville, Tenn.: Abengdon, 1976, in four vols., plus supplement): Supplement, p. 949.

9. Acts 2:2 (RSE).

10. Robert Graves, *The Greek Myths* (Baltimore, Md.: Penguin, 1955), 1:27–8.

11. Homer, *The Odyssey*, translated by Robert Fitzgerald. (New York: Doubleday, Anchor edition, 1963): 167; also see Gustav Schwab. *Gods and Heroes: Myths and Epics of Ancient Greece* (New York: Pantheon, 1946): 651–2. It is, of course, from Homer's Aeolus that the Aeolian science, the study of atmospheric winds, draws its name.

12. Geoffrey Irwin, *The Prehistoric Exploration and Colonization of the Pacific* (Cambridge: Cambridge Univ. Press, 1992): 5 and 25–30.

13. Ibid., 6 and 43–4.

14. Lionel Casson, *Ships and Seamanship in the Ancient World* (Princeton, N.J.: Princeton Univ. Press, 1971): 18–22, following page 370, fig. 6; also see Henry Hodges, *Technology in the Ancient World* (New York: Knopf, 1970): 95–6; and Rupert Sargent Holland, *Historic Ships* (Philadelphia: MaCrae, Smith, 1926): 19–23.

15. See Lionel Casson, *The Ancient Mariners* (New York: Macmillan, 1959).

16. Walter Prescott Webb, *The Great Frontier* (Boston: Houghton Mifflin, 1952).

17. Edward J. Kealey, *Harvesting the Air: Windmill Pioneers In Twelfth-Century England* (Berkeley: Univ. of California Press, 1987): 2.

18. K. D. White, *Greek and Roman Technology* (Ithaca, N.Y.: Cornell Univ. Press, 1984): 57.

19. Lynn White Jr., *Medieval Technology and Social Change* (Oxford: Oxford Univ. Press, 1962): 80; James Burke, *Connections* (Boston: Little, Brown, 1978): 85; Kealey, *Harvesting*, 3.

20. White, *Medieval Technology*, 86; Kealey, *Harvesting*, 10, footnote 9; Dennis G. Shepherd, "Evolution of the Modern Wind Turbine." Shepherd's ms., a copy of which is in the possession of the author, has been published as part of David A. Spera, ed., *Wind Turbine Technology* (Fairfield, N.J.: American Society of Mechanical Engineers, 1994).

21. Ahmad Y. al-Hassan, Donald Hill, *Islamic Technology* (New York: Cambridge Univ. Press, 1986): 55; Charles Singer et. al., eds., *A History of Technology* (Oxford: Clarendon, 1956, in five vols.): II, 615–6.

22. Lewis Mumford, *Technics and Civilization* (New York: Harcourt, Brace, 1934): 115.

23. See Kealey, *Harvesting*, 10–2; White, *Medieval Technology*, 87; Shepherd, "Evolution of the Modern Wind Turbine," 12–9.

24. Burke, *Connections*, 89.

25. Terry S. Reynolds, *Stronger than a Hundred Men: A History of the Vertical Wheel* (Baltimore: Johns Hopkins Univ. Press, 1983): 48.

26. Bertrand Gille, *The History of Techniques*, translated from French by P. Southgate and T. Williamson (New York: Gordon and Breach, 1986, in two vols.): I, 457. Also see Walter Minchinton, "Wind Power," and Walter Minchinton, Perveril Meigs, "Power From the Sea," both in *History Today* 30 (March 1980): 31–36, 42–46.

27. White, *Medieval Technology*, 85.

28. Gille, *The History of Techniques*, I, 460. It is instructive to note that Don Quixote's famous encounter with windmills on the plains of La Mancha was not with one or two but "thirty or forty lawless giants." See Miguel de Cervantes Saavedra, *Don Quixote de la Mancha*, translated by Samuel Putnam (New York: Viking, 1949, in two vols.): I, 62–3.

29. Jean-Claude Debier, Jean-Paul Deleage, Daniel Hemery, *In the Servitude of Power: Energy and Civilisation through the Ages*, translated by John Barzman (Atlantic Highlands, N.J.: Zed Books, 1991 edn.):78.

30. Lynn White, *Medieval Religion and Technology* (Berkeley: Univ. of California Press, 1978): 223.

31. Carolyn Merchant, *The Death of Nature: Women, Ecology, and the Scientific Revolution* (San Francisco: Harper & Row, 1980): 45.

32. Kealey, *Harvesting*, 107–153, passim.

33. See Donald Worster, *Rivers of Empire: Water, Aridity & The Growth of the American West* (New York: Pantheon, 1985), 22–31, for an enlightening exploration of Karl Wittfogel's controversial "hydraulic society" thesis.

34. Kealey, *Harvesting*, 50, 57, 69–70, 107–8.

35. Kealey explores the Abbot Samson/Herbert issue in great detail between pages 132 and 153.

36. Debeir, Deleage, and Hemery, *Servitude*, 78–9.

37. Ibid., 208–9. The idea of controlling power sources through ownership of water rights persisted well into the nineteenth century. So, however, did the egalitarian idea. A Scottish professor, speaking before the Glasgow Philosophical Society in 1888, remarked that he "preferred wind power to water, because the latter was usually tied up. The landlords, however, had not yet claimed a right over the wind." *Electrical World* 11 (June 23, 1888): 315.

38. T. K. Derry, Trevor I. Williams, *A Short History of Technology* (New York: Oxford Univ. Press, 1961): 257.

39. Shepherd, "Evolution" (ms.pages 24–5). In further praise of these Dutch windmills, it should be mentioned that some have remained operational for hundreds of years, whereas engineers figure the life of today's wind turbines at thirty years.

40. John Reynolds, *Windmills and Waterwheels* (New York: Praeger, 1975): 88. It is worth noting that controlling the driving force of a windmill is more difficult than operating a sailboat. The power of a sailboat is measured with the square of the wind, whereas in a windmill it goes with the cube. At a windspeed of 25 mph the miller has to contend with a force twenty-five times greater than does the sailor.

41. Merchant, *Death*, 149–163.

42. Ibid., 89.

43. Shepherd, "Evolution," 28; Volta Torrey, *Wind-Catchers: American Windmills of Yesterday and Today* (Brattleboro, Vt.: Stephen Greene, 1976): 46–56.

44. G. P. Tennyson, "Potential of Wind Energy Conversion Systems in the Great Plains," *Proceedings of the Solar and Wind Systems Workshop* (Lincoln, Nebr.: Great Plains Agricultural Council Energy Committee, 1983): 73.

45. Mumford, *Technics*, 117.

46. Ibid., 112–3.

47. Merchant, *Death*, p. 217.

48. Ibid., 118.

49. White, *Medieval Religion*, 22.

50. Louis C. Hunter, *A History of Industrial Power in the United States: Volume One, Waterpower* (Charlottesville: Univ. Press of Virginia, 1979): 3.

51. Henry Bronson, *History of Waterbury, Connecticut*, in two vols. (Waterbury, 1858): 29, as mentioned in Hunter, *Industrial Power*, 6–7.

52. See Terry S. Reynolds, *Stronger than a Hundred Men: A History of the Vertical Water Wheel* (Baltimore: Johns Hopkins Univ. Press, 1983): 311, 328–9).

53. Larry Hasse, "Watermills in the South: Rural Institutions Working Against Modernization," *Agricultural History* 58 (July, 1984): 283.

54. Hunter, *Industrial Power* 29; Hasse, "Watermills in the South," 283.

55. Volta Torrey, *Wind-Catchers: American Windmills of Yesterday and Today* (Brattleboro, Vt.: Stephen Greene, 1976): 37–40. Local artisans reconstructed the Robertson mill and it is in operation today.

56. John W. Oliver, *History of American Technology* (New York: Ronald, 1956): 27.

57. As quoted in Leo Marx, *The Machine in the Garden* (New York: Oxford Univ. Press, 1964): 153.

58. There is no reliable evidence that the wind can cause mental illness, insanity, or suicide. However, it is a widely held belief.On the topic of such mental illness, see Dorothy Scarborough's novel, *The Wind* (New York: Grosset & Dunlap, 1925). Written in the mid-1920s, it describes the shattering effect of a West Texas environment on a young, Southern-born bride as she struggles, unsuccessfully, to cope with a hostile world.

59. Wister to his mother, July 28, 1885, in Francis K. W. Stokes, *My Father, Owen Wister* (Laramie, Wyo., 1952): 40. Wister, for his part, was referring to the Wyoming country.

60. See David Dary, *True Tales of Old-Time Kansas* (Lawrence, Kans.: Univ. of Kansas Press, 1985).

61. Ibid., 40–7, *passim*. It is difficult to comprehend how Peppard could sail successfully to the West against the prevailing northwest winds of the Plains. Given the construction and rigging of his wind wagon, the technique of tacking against the wind would seem impossible, but he must have devised a way.

62. Walter Prescott Webb, *The Great Plains* (New York: Grosset & Dunlop, 1971 edn.): 337–47. Also see A. Bower Sageser, "Windmill and Pump Irrigation on the Great Plains: 1890–1910," *Nebraska History*, 48 (summer, 1967): 107–18; Everett Dick, "Water: A Frontier Problem," *Nebraska History*, 49 (autumn, 1968): 215–245; Terry G. Jordan, "Windmills in Texas," *Agricultural History* 37 (April, 1963): 80–5; T. Lindsay Baker, "Turbine-Type Windmills of the Great Plains and Midwest," *Agricultural History* 54 (January, 1980): 38–51. With the publication of Baker's *Field Guide* in 1985, we have the definitive history of the water-pumping American windmill.

63. Although historians generally name Halladay as first inventor of an American windmill, there is a difference of opinion. Roger S. Manning maintains that "there is evidence to suggest that a form of self-regulating windmill was developed in California as early as 1849." By 1861 hundreds of these locally built units could be seen on the west side of the Central Valley. See Roger S. Manning, "The Windmill In California," *Journal of the West* 14 (July, 1975): 33–39.

64. Dennie Landt, *Catch the Wind: A Book of Windmills and Windpower* (New York: Four Winds Press, 1976): 61; *Kansas Wind Energy Handbook* (Kansas Energy Office, 1981): 2; Volta Torrey, "Catching the Western Winds," *American Heritage* 20 (March/April, 1983): 48.

65. Baker's *Field Guide* presents the types and brands in exceptional detail.

66. Dick, "Water: A Frontier Problem," 242.

67. *Scientific American* 49 (October 13, 1883): 227. The *arrastre*, a machine of Spanish and Mexican origin was used to reduce ore in dredging and placer mining.

68. Joan Baeza, "Where the Wind Blows Free," *Arizona Highways Magazine* 57 (August, 1981): 2.

69. Jordan, "Windmills in Texas," 85.

70. Larry D. Hodge, "Gone with the Windmills," *Texas Highways* (December, 1988): 11.

71. Jordan, "Windmills in Texas," 85.

72. Quoted in Stanley Freese, *Windmills and Millwrighting* (Devon, England: David & Charles Newton Abbot, 1971): xiv.

73. John R. Erickson, "Windmills: Major Source of Tomorrow's Energy," *Oklahoma Today* (autumn, 1978): 5.

74. Marx, *Machine In the Garden*, 4.

75. See Bill George, "Reaping the Wind," *American Heritage of Invention & Technology* 8 (winter, 1993):14.

76. Torrey, *Wind-Catchers*, 148–50; T. Lindsay Baker, *A Field Guide to American Windmills* (Norman: Univ. of Oklahoma Press, 1985):4.

77. See Yi-Fu Tuan, *Topophilia: A Study of Environmental Perception, Attitudes, and Values* (New York: Columbia Univ. Press, 1974).

78. The issue of perception will be explored in greater depth in later chapters. It is a central issue, and I believe that if electricity is ever to be produced by wind technologies in significant amounts, gaining an understanding of symbolic and perceptual attitudes will be every bit as important as solving the technical problems.

CHAPTER 2. A CONNECTION IS MADE: WIND INTO ELECTRICITY

1. Quoted in Bern Dibner, *Benjamin Franklin: Electrician* (Norwalk, Conn.: Burndy Library, 1976): 16.

2. Malcolm MacLaren, *The Rise of the Electrical Industry During the Nineteenth Century* (Princeton, N.J.: Princeton Univ. Press, 1943). See pages 41, 46.

3. Quoted in Marc Bloch, *The Historian's Craft* (New York: Knopf, 1959): 66. The famous French historian cited Valéry's comment as an example of a "vast subject which has still received no serious treatment."

4. MacLaren, *Rise of Electrical*, vi.

5. See Sam H. Schurr and Bruce C. Netschert, *Energy and the American Economy, 1850–1975* (Baltimore, 1960); Martin V. Melosi, *Coping With Abundance: Energy and Environment in Industrial America* (Philadelphia: Temple Univ. Press, 1985): 52–67.

6. Charles Singer, et al., *A History of Technology* (Oxford: Clarendon, 1958, in five vols.): V, 230–1.

7. A primary problem with hydroelectricity in the nineteenth century was storage. Engineers had not yet devised massive poured concrete dams that are so prevalent today.

8. Schurr and Netschert, *Energy*, 54, 485–7.

9. Ibid., 32.

10. Howard Mumford Jones, *The Age of Energy* (New York: Viking, 1970); Amory Lovins, *Soft Energy Paths* (New York: Ballentine, 1977).

11. Schurr and Netschert, *Energy*, 55.

12. For a discussion of these phenomena see Robert Thayer Jr., *Grey World, Green Heart* (New York: John Wiley, 1994). Leo Marx in *Machine in the Garden* is convincing in his thesis that the duality between technology and the pastorale were evident from the beginning of the republic.

13. William Thompson, "The Sources of Energy in Nature," *Engineering* 32 (London, September 23, 1881): 321–2. Also see Crosbie Smith and M. Norton Wise, *Energy and Empire: A Biographical Study of Lord Kelvin* (New York: Cambridge Univ. Press, 1989), and Silvanus P. Thompson, *The Life of William Thompson* (London: MacMillan, 1910, in two vols.), I, 289–91.

14. Thompson, "The Sources of Energy in Nature," 322; Smith and Wise, *Energy and Empire*, 658.

15. Smith and Wise, *Energy and Empire*, 714.

16. Thompson, "The Sources of Energy."

17. See Carolyn Marvin, *When Old Technologies Were New* (New York: Oxford Univ. Press, 1988).

18. *Scientific American* 50, 4 (January 26, 1884): 53.

19. The articles in the following list give a sampling of new windmills and modifications of old ones described and recorded in the *Scientific American*: "Improved Windmill," vol. 32 (April 10, 1875): 230; "Smith's Improved Windmill," vol. 38 (February 9, 1878): 83; "Improved Windmill," vol. 42 (May 8, 1880): 296; "Improved Wind Motor," vol. 44 (June 18, 1881): 390; vol. 48 (March 3, 1883: 130; "A New Wind Engine," vol. 50 (March 29, 1884): 198; "Improved Windmill," vol. 55 (July 24, 1886): 50; "An Improved Windmill," vol. 59 (October 6, 1988): 217.

20. The *Kansas Wind Energy Handbook* (Kansas Energy Office, 1981), on page 3, states that Farmer patented a wind-electric device in 1860. I have not been able to confirm this, although there is no reason to question it. Farmer had the knowledge and the ability to invent such a machine. However, according to one source, a Belgian professor may have predated Farmer's accomplishment by some twenty years. A. M. Tanner wrote in 1892 that Professor François Nollet of Brussels patented "the idea of generating electricity through natural forces, namely wind and water" on January 6, 1841. Patenting an *idea*, of course, might have meant very little and would not be accepted by the U.S. Patent Office. See A. M. Tanner, "The Electrical Utilization of Water and Wind Power First Proposed by Nollet in the Year 1840," *Electrical World* 19 (April 9, 1892): 242.

21. Moses Farmer has not been the subject of a biography; however, considerable information can be found in "Moses G. Farmer," *Dictionary of American Biography* (1931), VI, 279–81; "Moses Garrish Farmer," *The National Cyclopaedia of American Biography* (New York: James T. White, 1897), VII, 361–2; "Moses Farmer," *Harper's Encyclopaedia of United States History* (New York: Harper & Brothers, 1902, in ten vols.), III, 311.

22. H. W. Pond, "Heat Without Coal—Utilization of Wind Power," *Scientific American* 18 (January 4, 1868):3.

23. "Wind Wheels—Some of Their Adaptations," *Scientific American* 18 (June 13, 1868):377.

24. "The Storage of Wind Power," *Scientific American* 49 (July 14, 1883d):17.

25. Ibid.

26. "Storing Wind Power For Small Motors," *Scientific American* 49 (December 8, 1883):353.

27. Letter from F. A. R., of Brewster, Mass., in *Scientific American* 49 (November 10, 1883):293.

28. Correspondence from F. N. Davis of Calais, Maine, *Scientific American* 49 (August 11, 1883):148.

29. "Storing Wind Power! Revised and Corrected," *Scientific American* 49 (September 8, 1883):148.

30. Ibid.

31. Letter from C. C. R., *Scientific American* 49 (October 13, 1883):229. This idea of breaking down water into hydrogen and oxygen has been proposed in contemporary times.

32. Letter from J. P. M., *Scientific American* 49 (September 22, 1883):181. An earlier illustrated article, titled "Improved Wind Wheel and Water Elevator," *Scientific American* 30 (June 27, 1874):402, showed a substantial wind wheel with water buckets, but no storage system.

33. See Thomas P. Hughes, *American Genesis: A Century of Invention and Technology* (New York: Viking, 1989):1–13.

34. Richard H. Schallenberg, *Bottled Energy: Electrical Engineering and the Evolution of Chemical Energy Storage* (Philadelphia: American Philosophical Society, 1982): 52–68, passim.

35. Much of the information and primary material was obtained from the C. F. Brush Collection, Case Archives of Contemporary Science and Technology, Freiberger Library, Case Western Reserve University, Cleveland (Hereafter, Brush Papers).

36. "Mr. Brush's Windmill Dynamo," *Scientific American* 63 (December 20, 1890): 389.

37. Mel Gorman, "Charles F. Brush and the First Public Electric Street Lighting System in America," *Ohio Historical Quarterly* 70 (April, 1971): 129.

38. Charles F. Brush, "The Arc-Light," *Century Magazine* 70 (1905): 112.

39. Ibid.

40. "Charles Francis Brush," *The National Cyclopaedia of American Biography*, V. 21, 1.

41. Telephone interview with Baldwin Sawyer conducted by the author, October 30, 1988. Baldwin Sawyer, son of Charles Baldwin Sawyer, who was perhaps Brush's closest confidant, recollects that his father often spoke of Brush's fascination with solar energy and how to transform it for the use of human beings.

42. See "Charles Francis Brush," *Dictionary of American Biography*, V. 11, Supp. I, 129.

43. Brush, "The Arc-Light," 115; Gorman, "Charles F. Brush," passim.

44. Brush Electric Machines, Ltd., exists today as a subsidiary of the Hawker Siddeley Group, primarily operating in England as a manufacturer of electric motors of worldwide reputation. The company has provided generators for James

Howden & Co., a contemporary Scottish manufacturer of wind turbines. Thus, the Brush tradition in wind harnessing continues.

With regard to the Brush arc light, historian MacLaren argues that Brush made a great contribution for "electric illumination which allows in large measure for this phenomenal growth." MacLaren, *Rise of Electrical* 69–70.

45. Margaret Emily Richardson manuscript, ch. 12, page 9, in Brush Papers, Box 25.

46. "A Great American Inventor." Page 5 of a twelve-page manuscript, dated January 15, 1930, in Box 20, Brush Papers.

47. In fact, I found nothing at all in the Brush papers regarding the wind generator. It is inconceivable that Brush, with his penchant for scientific method, failed to keep records on the performance of the generator, but the material did not survive—an indeed unfortunate turn of events.

48. The manuscript, "A Great American Inventor," states that the only electricity before Brush's mansion was J. Pierpont Morgan's in New York. However, this appears to be highly unlikely. The Pearl Street central electrical generating plant opened in New York in 1882, powering thousands of incandescent lights; surely, a few of these lights were installed in private homes. See Thomas Hughes, *Networks of Power* (Baltimore: Johns Hopkins Univ. Press, 1983):40–45.

49. An article in the Cleveland *Plain-Dealer*, March 14, 1930, stated that blueprints for the wind dynamo existed, but nothing is available in archival material.

50. See "Storing Wind Power, Revised and Corrected," *Scientific American* 49 (September 8, 1883): 148, and "The Storage of Wind Power," *Scientific American* 49 (July 14, 1883):17.

51. "Mr. Brush's Windmill Dynamo," *Scientific American* 63 (December 20, 1890): 389.

52. Ibid.

53. Schallenberg, *Bottled Energy*, 53–4.

54. "Mr. Brush's Windmill Dynamo," *Scientific American*, 389.

55. The manuscript, "A Great American Inventor." The typescript is signed C. S. Sawyer.

56. Ibid.

57. See Hughes, *Networks of Power*, for the best analysis of the worldwide effort to centralize and consolidate electric power networks.

58. For more on this thrust toward complex electrical systems see Hughes, *Networks of Power*, and the 1989 work by Hughes entitled, *American Genesis*.

59. "Mr. Brush's Windmill Dynamo," Scientific American, 389.

60. Ibid.

61. "Charles Francis Brush," page 5 of ms. by Marie Gilchrist, in Box 20, Brush Papers.

62. Cleveland *Plain-Dealer*, June 16, 1929.

63. Cleveland *Plain-Dealer*, October 19, 1929.

64. Harry Eisenman III, "Charles F. Brush: Pioneer Innovator in Electrical Technology," (Ph. D. dissertation, Case Institute of Technology, 1967): 155–178, passim. This work is of limited value. The Brush wind dynamo is not mentioned.

65. The headline for the eulogy on Brush in the Cleveland *Plain-Dealer*, June 16, 1929, read: "Half Century of Service Ends for Inventor-Philanthropist." Also see David D. Van Tassel, John J. Grabowski, eds., *The Encyclopedia of Cleveland History* (Bloomington, Ind.: Indiana Univ. Press, 1988): 133, 369–70, 953.

66. Charles Brush to James Bishop, April 4, 1929, quoted on page 13, ch. 12, of Margaret E. Richardson's draft manuscript for a biography of Charles Brush, in Box 25, Brush Papers.

67. See Nona H. Schirg, Brush's secretary to James W. Bishop, December 12, 1929, in Box 4, Brush Papers.

68. See Charles B. Sawyer to James W. Bishop, November 18, 1930, in Box 4, ibid.

69. Cleveland *Plain-Dealer*, March 14, 1930.

70. C. B. Sawyer to Professor Elihu Thomson, April 15, 1930, in Box 4, Brush Papers.

71. Page 14 of Richardson's draft in Box 25, Brush Papers.

72. Bishop to Dr. Roger G. Perkins, August 7, 1930, in Box 4, Brush Papers.

73. Roger Perkins to Bishop, August 25, 1930, ibid.

74. Bishop to Schirg, November 26, 1930, ibid.

75. William Wickenden, president, to Alexander Brown, January 22, 1930, in Charles F. Brush Papers, 1930–1, Western Reserve Historical Society, Cleveland.

76. Brush Memorial Fund flyer, ibid.

77. Brown to Van Sweringen, July 14, 1930, ibid.

78. See Brown to Randolph Eide, president of the Ohio Bell Telephone Co., November 11, 1931, and Charles B. Sawyer to Roger G. Perkins, November 4, 1931, ibid.

79. Brush's wife died a good many years before he did. They had three children. In his will, Brush specified that if his immediate heirs did not wish to live in the mansion—and they did not—it should be torn down. The mansion was demolished to make room for commercial buildings.

80. The story of the windmill's end was related to the author by Baldwin Sawyer, September 22, 1988. Some photographs of the windmill apparatus, taken in the 1950s and can be found in the Brush Papers, 1930–1, Western Reserve Historical Society, Cleveland.

CHAPTER 3. EXPLORING THE POSSIBILITIES

1. "Harnessing Nature," *Scientific American* 108 (April 5, 1913):304. The editors noted that this "cheerful submission" of nature did not include earthquakes, particularly the 1906 event in San Francisco.

2. See Peter J. Schmitt, *Back To Nature: The Arcadian Myth In Urban America* (Baltimore: Johns Hopkins Univ. Press, 1969; 1990, reprint):3–19, passim; Also T. J. Jackson Leer, *No Place of Grace: Antimodernism and the Transformation of American Culture, 1880–1920* (New York: Pantheon, 1981).

3. H. E. M. Kensit. "Electric Power From Wind," *Canadian Engineer* 21 (December 14, 1911):671.

4. "An Electric Plant Operated by an Air Turbine," *Scientific American* 102 (January 22, 1910):86.

5. Charles B. Hayward, "Powerful German Windmills," *Scientific American* 92 (March 25, 1905):245–6; Also "The Electrical Value of Wind Power," *Scientific American* 93 (November 18, 1905):394–5.

6. "Remarkable Wind and Sun Power Plant," *The Implement and Machinery Review* 36 (London, October 1, 1910):722–3; and R. A. Fessenden, "Utilizing the Energy of Solar Radiation, the Wind, and Other Intermittent Natural Sources," *Mechanical Engineer* no. 18569 (Manchester, U.K., November 4, 1910). A contemporary, A. H. Gibson, challenged Fessenden's mathematics. Gibson contended that a rough estimate would show "that the cost of such a storage plant to store a fortnight's supply [of electricity] would be in the neighborhood of £140 per horse-power, and that the annual charges and depreciation would amount to some £26 per horse-power." Even with a free source of energy, the cost of utilizing the wind "would be enormously in excess of the cost of steam, gas, or water power." A. H. Gibson, *Natural Sources of Energy* (London: Cambridge Univ. Press, 1913): 117–8.

7. Hayward, "Powerful German Windmills":245.

8. "The Wind Mill Industry," *Farm Implement News* 15 (Chicago, December 6, 1894):20.

9. Letter to the editor, signed Holt, in *Scientific American* 66 (February 27, 1892):133.

10. Clark Spence, "Early Uses of Electricity in American Agriculture," *Technology and Culture* 3 (spring, 1962):152. Spence quotes the editor of the *Rural New-Yorker* 47 (June 9, 1888):386.

11. "Synopsis, Report of the Committee on Electricity on the Farm," National Electric Light Association, *Papers, Reports and Discussions* (36th convention, June 2–6, 1913), Commercial Sessions volume, xi. As cited in Spence, "Early Uses," 154.

12. See James Williams, "Energy Resources and Uses in Rural California," Univ. of California, Santa Barbara, Ph.D. dissertation in history, 1984.

13. Historian Donald Worster focuses on farmers' dependence upon huge federal water distribution systems in *Rivers of Empire* (New York: Pantheon Books, 1985).

14. Quoted by William E. Smythe in *The Conquest of the Arid West* (New York: Harper & Row, 1900):111.

15. Entrepreneur Samuel Insull successfully applied the secret of systems to electricity, lacing the United States with power lines. Insull's role in rural electrification is discussed in chapter 5. See Thomas Hughes, *American Genesis*.

16. *Scientific American* 108 (April 5, 1913):302.

17. Information on the Lewis Electric Company was obtained from a trade brochure in the possession of historian T. Lindsay Baker of Baylor University, who generously shared its contents with the author. The original catalogue (circa late 1893 or 1894) promoted graphite bushings, products of the Graphite Lubricating Company, that were used in dynamos built by the Lewis Electric

Company. No records have survived to indicate how many Lewis units might have been sold.

18. This account is based primarily on an article in *Electrical Engineer* 18 (November 21, 1894):412–3. Also see "Illustrated Catalogue of Windmills, Tanks, and Pumps as Applied to Water Supply Systems, also Windmills as Adapted for Power," catalogue for Charles J. Jager Company, circa 1895, found in "Windmill Catalogues and Literature," PanHandle Plains Historical Museum, Canyon, Texas.

19. Quote from brochure in the possession of T. Lindsay Baker.

20. T. Lindsay Baker, *A Field Guide to American Windmills* (Norman: Univ. of Oklahoma Press, 1985):45–6, 210.

21. "The Wind Mill Industry," *Farm Implement News* 15.

22. "La Verne Noyes: A Biographical Sketch," typescript, in "Aermotor, Folder 101," Windmill Literature, Panhandle-Plains Historical Museum.

23. See "The Future of the Windmill," *Scientific American* 108 (April 5, 1913):309; and Putnam A. Bates, "Light and Power on the Farm," *Scientific American* 104 (April 15, 1911):380.

24. See Forrest McDonald, *Insull* (Chicago: Univ. of Chicago Press, 1962), and Hughes, *American Genesis*.

25. Theodore Roosevelt "Opening Address of the President," in Newton C. Blanchard, ed., *Proceedings of a Conference of Governors in the White House* (Washington, DC: U.S. Government Printing Office, 1909):5–10.

CHAPTER 4. WIND ENERGY FINDS A PLACE

1. E. N. Fales, "A New Propeller-Type, High-Speed Windmill for Electric Generation," American Society of Mechanical Engineers, *Transactions* 49–50 (1927–8): 1.

2. N. G. Calvery, *Windpower Principles: Their Application on the Small Scale* (London: Charles Griffin, 1979.): 11.

3. Fales, "A New Propeller-Type," 1.

4. Ibid.

5. Ibid.

6. Ibid. While Fales was making progress with new wind turbines, Juan de la Cierva, a Spanish inventor, was designing, and flying (1923), an "autogyro," the predecessor to the helicopter. Both Fales and de la Cierva made notable contributions to work on rotors. See Alfred Gessow and Garry C. Myers Jr., *Aerodynamics of the Helicopter* (New York: Frederick Ungar, 1952, reprinted 1967): 13–5.

7. Waldemar Kaempffert. *Modern Wonder Works: A Popular History of American Invention* (New York: Scribner's, 1924): 504.

8. Ibid.: 534. Although Kaempffert evidently knew nothing of Brush's wind dynamo, on page 563, ibid., he gives credit to Brush for the invention of the arc light system.

9. R. E. Ruggles, "Colorado Wind Supplies Electric Light," *American Thresherman and Farm Power* 25 (August 1922): 6.

10. "Transforming Wind Into Energy," *Dakota-Farmer* 41 (February 15, 1921):239. This article does not mention the Aerolite Wind Electric Company. However, two letters (George Manikowski, president, Aerolite Wind Electric Company, to Woodmanse Mfg. Co., both dated September 24, 1921) in folder 116, mss. 1292, Fritchle Wind-power Electric Company papers, Colorado Historical Society, Denver (hereinafter Fritchle papers), connect the Manikowski brothers with Aerolite.

11. Ruggles, "Colorado Wind," 6.

12. U.S. Department of Agriculture. *Year Book, 1921*: Table 417, page 788.

13. F. E. Powell. *Windmills and Wind Motors: How To Build and Run Them* (New York: Spon & Chamberlain, 1918): iii.

14. Ibid., 66.

15. Ibid., iii.

16. Oliver Fritchle to George Derby, editor, James T. White Company, November 15, 1921, in folder 116, mss. 1292, Fritchle papers. Other wind-energy company papers generally have not survived.

17. Aermotor Company to O. P. Fritchle, July 20, 1917, and July 28, 1917; G. Trindle, Flint & Walling Mfg. Co., to Fritchle Electric Company, July 20, 1917, July 27, 1917; H. Holland, Perkins Wind Mill Company to the Fritchle Electric Company, July 27, 1917; W. Johnson, the Elgin Wind Power and Pump Company to O. P. Fritchle, July 18, 1917, in folder 13, Fritchle papers.

18. Ibid., Aermotor Company to O. P. Fritchle, July 20, 1917.

19. Fritchle visited the Perkins Wind Mill Company in October, 1921. To an editor he reported that Perkins was still experimenting, having produced only two plants. He judged their experimental plants inferior: they did not cut off or switch on automatically. Fritchle perhaps misread the intent of Perkins, for in 1922 the Aerolectric was in full production. O. P. Fritchle to George Derby, editor, White & Co., November 15, 1921, in folder 116, Fritchle papers.

20. "Fritchle Wind Mill owners by State," in folder 118, Fritchle papers.

21. Henry Anderson to O. P. Fritchle, undated, in folder 122, Fritchle papers. All windmill companies took a great interest in obtaining letters of praise for their products.

22. See correspondence in folders 116, Fritchle papers.

23. The Fritchle papers do not reveal existence of a contract or what the arrangement was between Fritchle and Woodmanse.

24. Most of the letters to Henderson are in folder 121, Fritchle papers. Stan Oliner, Colorado Historical Society curator, discovered the Henderson letter ruse and pointed it out to the author.

25. T. Lindsay Baker, *A Field Guide to American Windmills* (Norman: Univ. of Oklahoma Press, 1985): 284–7.

26. The description is taken from sales material in Perkins, folder 2, windmill lit. collection, Panhandle-Plains Historical Museum, Canyon, Texas.

27. Ibid., folders 1 and 2.

28. Katherine Kay Jellison, *Entitled To Power: Farm Women and Technology, 1913–1963* (Chapel Hill: Univ. of North Carolina Press, 1993): 42.

29. Perkins, folder 4, Panhandle-Plains Historical Museum.

30. C. A. Carlisle Jr., "Converting Wind to Electricity," *American Thresherman* 24, no. 8 (February, 1922).

31. Calvary, *Windpower Principles*, 11.

32. Bucklen folder, windmill lit. collection, Panhandle-Plains Historical Museum. This advertising was put out about 1930; hence, the reference to Lindbergh's 1927 flight. HEBCO had begun producing machines in 1921 or 1922.

33. Ibid.

34. Publicity flyer, circa 1922, ibid.

35. Ibid.

36. See Alan L. Olmsted, Paul Rhode, "The Farm Energy Crisis of 1920," *Agricultural History* 62 (winter, 1988):48–60. The fuel shortage resulted from cutbacks by suppliers. Many believed this to be a plot by the companies, to raise prices, since, in general, gasoline was abundant in the 1920s.

37. Wendell Berry, *The Gift of Good Land* (San Francisco: North Point, 1981):128-31.

38. Wallace to Taylor, August 15, 1923; and memorandum from MacDonald, July 15, 1924, in correspondence to secretary, 1923 and 1924, acc. 234, dr. 234 & 440, Farm Power folder, USDA, record group 16, National Archives, Washington DC.

39. Memo to the secretary from MacDonald, July 15, 1924, ibid.

40. William Aitkenhead, "Electric Windmills," *Wallaces' Farmer* 50 (December 4, 1925).

41. "A Report on the Use of Windmills for the Generation of Electricity," Bulletin 1, Institute of Agricultural Engineering, Univ. of Oxford, (Oxford: Clarendon, 1926): 8.

42. "Windmill-Driven Electric Generators," *Engineering*, 122 (London, July 30, 1926):150. This article is a report on the Harpenden experiment. For other reports, see "The Growing Popularity of the Modern Windmill," T*he Implement and Machinery Review* 51 (London, August 1, 1925):425–6; "Windmill Generating Plants," *The Implement and Machinery Review* 52 (March 1, 1926):1220–24; "Electricity Generation by Windmills," *The Implement and Machinery Review* 52 (June 1, 1926):181.

43. The British experiment results were very popular. The first editionn (1926) sold out within two years. In 1933, as a result of demand, a summary report was published. See C. A. Cameron Brown, "Windmills for the Generation of Electricity," Inst. for Research in Agricultural Engineering (Oxford, January, 1933). The wind machines were all of British or European origin. The seven machines were: the Airolite, Grier & Mackay Ltd., Glasgow; Garty-Apex (two machines), Glasgow Electrical Engineering Co.; Ventimotor (two machines), Patent Lighting Co.; Aerodynamo, by Aerodynamos, Ltd., London; Agricco by Osmond & Young, London. The Aerodynamo and the Agricco were I believe, built in Europe.

44. William Aitkenhead, "Electric Windmills: Getting Your Power From the Wind," *Wallaces' Farmer* 50 (December 4, 1925):33.

45. E. W. Lehmann, "Electric Power From the Wind," *Successful Farming* 28 (October, 1929):18, 98. Lehmann admitted that testing at Purdue University and Iowa State College indicated that the wind was insufficient for satisfactory performance in summer.

46. James Clebourne, "Reaping the Wind," *Palm Springs Villager Magazine* (April, 1957):13.

47. Paul W. Travis, "The Wind Machine in the Pass," *Westways* 48 (February, 1956): 12 (12–3).

48. "Electrical Current From the Wind," *Compressed Air Magazine* 33 (April, 1928):2380.

49. Ibid., 13.

50. James Clebourne, "Reaping the Wind," 13.

51. Ibid. Also see Walter Ford, "Power From the Wind," *Desert* (June 1975): 22–3.

52. Tom Patterson, "Out of the County's Past," Palm Springs *Enterprise*, January 12, 1979, B-4; Clebourne, "Reaping the Wind," p. 13.

53. Although the Oliver machine was dismantled, the foundation remains. It is approximately one-quarter mile NW from the rest stop on the north lanes of Interstate 10 in San Gorgonio Pass. Information from an interview with Robert Scheffler, Southern California Edison, November 19, 1992; and letter from Shefrah Ann Rozenstain to Nancy Robinson, reference librarian, Palm Springs Library, May 21, 1989, in clipping file, Wind Energy, Palm Springs Library.

54. "Interview with Marcellus Jacobs," *Mother Earth News* 24 (November, 1973):53.

55. Ibid.

56. For a description of the development of the Jacobs unit, see Marcellus L. Jacobs, "Experience with Jacobs Wind-Driven Electric Generating Plants, 1931–1957," in *Proceedings of the United Nations Conference on New Sources of Energy, Wind Power*, vol 7., Solar Energy, Wind Power and Geothermal Energy, Rome, August 21–31, 1961: 337–9.

57. Paul Jacobs, president, Jacobs Wind Electric Co., to author, May 10, 1993.

58. This paragraph is based on the interview with Jacobs, *Mother Earth News* 24, pages 55 to 57, and an interview with Marcellus Jacobs's wife and their son, Paul, by the author, June 6, 1987. Also see Tom Kovarik, Charles Pipher, and John Hurst, *Wind Energy* (Chicago: Domus, 1979) page 12, for material on Marcellus Jacobs.

59. Interview with Bruns by the author, June 5, 1987. At the time of the interview, Bruns still had two Jacobs machines providing power for his South Dakota farmhouse. Bruns operated two systems. If the Jacobs 32-volt system faded he could switch to 110 volt high line central power.

60. Ibid.

61. Interview with Paul Jacobs, June 6, 1987.

62. Jacobs folder, windmill lit. coll., Panhandle-Plains Historical Museum.

63. "Statement of M. L. Jacobs," in U.S., Congress, House of Representatives, *Hearing on Wind Energy Before the Subcommittee on Energy of the Committee on Science and Astronautics*, May 21, 1974, 93rd Cong., 2nd Sess., p. 206.

64. Richard E. Byrd Jr. to M. L. Jacobs, June 17, 1955, ibid., 208.

65. Interview with Paul Jacobs, June 6, 1987.

66. Kovarik, et al., *Wind Energy*, p. 12.

67. In talking with older ranchers in eastern Wyoming it was clear to the author that many wind generators were operating in the 1940s and 1950s; it is just as evident that the wealthier ranchers owned the wind generators.

68. See Robert Caro, *The Years of Lyndon Johnson: The Path to Power* (New York: Knopf, 1982): 502–15, for a moving description of the hardships of the preelectric period in Lyndon Johnson's "Hill Country." It was a life "filled with drudgery."

69. Fred W. Hawthorn, "Farm Experience with Wind Electric Plants," *Agricultural Engineering* 19 (January, 1938):7-8.

70. Ibid., 8. The customer-per-mile factor referred to the REA requirement of three hookups per mile of high line.

71. Ibid., 7. The caption reads: "The Wind Electric Plant on the Author's Farm."

72. Ibid., 8.

73. Interview with Paul Jacobs, June 6, 1987.

74. Interview with Mrs. Marcellus Jacobs, June 6, 1987.

75. Interview with Marcellus Jacobs, *Mother Earth News*, 24 (November, 1973):57.

76. In Jacobs, "Experience With Wind-Driven," 338, Jacobs states that a 2500W plant cost $490. With a storage battery at $365 and a tower at $175, this gave a total of $1,030. Shipping and installation were extra, but required only two men's labor for two days and a little cement. Instructions were clear, and although the company had shipped hundreds of plants throughout the world it had not had "a single request for additional information to enable [owners] to erect the plant."

77. Interview with Paul Jacobs, June 6, 1987: Paul Gipe and Carl Judy, "Wind Chargers: Building Tools From the Nation's Past," *Mother Earth News* (April, 1983):116. Testifying before a congressional committee in 1945, R. F. Weinig, vice president and general manager of Wincharger Corporation, estimated that over 25,000 of the large plants were in operation in the United States. See U.S. House of Representatives. *Hearings on Rural Electrification Planning, H. R. 1742, Before a Subcommittee of the Committee on Interstate and Foreign Commerce*, June 13, 14, 15: October 16, 17, 18, 19, 30: November 2, 1945: 114.

78. Interview with Bruns, June 5, 1987.

79. Sears Roebuck & Co., Catalog, spring-fall, 1946: 645.

80. Interview with Mr. and Mrs. W. H. Jones, August 15, 1972, in oral history, Southwest Collection, Texas Tech Univ.: See Reynold M. Nik, "The Radio in Rural America During the 1920s," *Agricultural History* 55 (October, 1981):339–50.

81. House of Representatives, *Hearings on H. R. 1742*.

82. In a search for old wind plants to restore, Michael A. Hacklemann encountered these makes; he felt certain there were more. Hacklemann, *The Homebuilt, Wind-Generated Electricity Handbook* (Mariposa, California: Earthmind, 1975).

83. Cynthia Gregary, "Pioneer Recalls Wind Charger," *Plainview Daily Herald*, June 8, 1980, clipping in Southwest Collection, Energy-Wind, Texas Tech Univ.; also Interview with G. C. Applewhite by Richard Mason, October 15, 1980, in oral history, Southwest Collection, Texas Tech Univ.

84. L. Lothrope, "Wind Electrics," *Nor'-West Farmer* 52 (July, 1933):7.

85. Steve Hicks, "Wincharger and Jacobs," *Home Power* 11 (June/July, 1989): 13.

86. James B. DeKorne, "The Answer Is Blowin' in the Wind," *Mother Earth News* 24 (November, 1973):70 (67–75).

87. Interview with Bruns, June 5, 1987.

88. R. F. Weinig, testimony in U.S. House of Representatives, *Hearings on H. R. 1742*, 119.

89. Interview with Ida Chambers by author, June 30, 1986. Ida Chambers was in her nineties at the time of the interview. She has since passed away.

90. Interview with Becky Chambers by author, June 30, 1986.

91. Interview with Marcellus Jacobs, *Mother Earth News*, 58.

92. "Statement of Mr. M. L. Jacobs," in U.S. House of Representatives, *Hearings on Wind Energy Before the Subcommittee on Energy of the Committee on Science and Astronautics*, May 21, 1974, 93 Cong., 2nd Sess.: 206.

CHAPTER 5. THE REA OFFERS A DIFFERENT WAY.

1. See Thomas Hughes, *American Genesis* (New York: Viking, 1989).

2. See Hughes, *American Genesis*, 226–43 for a stimulating discussion of Insull's pivotal role in the development of U.S. electrical systems. Also Harold L. Platt, *The Electric City: Energy and Growth of the Chicago Area, 1880–1930.* (Chicago: Univ. of Chicago Press, 1991):275.

3. This attitude was again prevalent in the late 1970s and early 1980s with the passage of the Public Utility Regulatory Policy Act (PURPA), the scheme that forced utility companies to accept into their systems privately generated power.

4. Forrest McDonald, *Insull* (Chicago: Univ. of Chicago Press, 1962): 139–41. McDonald's book is a sympathetic appraisal of Insull.

5. Ibid., 141.

6. *The Electric Power Industry: Past, Present and Future*, by the editorial staff of the trade magazine, *Electrical World* (New York: McGraw-Hill, Inc., 1949):6. This work is a sympathetic history of the development of the power industry, published on the seventy-fifth anniversary of the magazine.

7. *Proceedings of the National Electric Light Association* 81, page 62.

8. Frank App, "The Industrialization of Agriculture," in *The Annals of the American Academy of Political and Social Science* 142 (March 1929):234.

9. *Harvests and Highlines* (Chicago: Middle West Utilities Co., 1930): 14–5, 57. This book—ghostwritten for Samuel Insull and his younger brother Martin J. Insull—explains the efforts of private utilities to provide for farmers' needs. It emphasizes that the successful, modern farmer needs electricity. For instance, on page 99 we need that "electric power has taught hens to lay higher priced eggs in the winter months. Pigs in electrified piggeries are not just pigs. The prophylactic pig in such a glorified sty is something entirely different from the proletarian pig in the oldtime wallow." Most significantly, the book provides an explanation for the failure of private power to extend central power to rural areas. The rationale is almost all economic. Also see Platt, *Electric City*, 269–70, and D. Clayton Brown, *Electricity for Rural America: The Fight for the REA* (Westport, Conn.: Greenwood, 1980):xv. The census of 1920 indicated that of the 6,000,000 farms in the United States only 452,620 had electric lights.

10. John M. Carmody, "Rural Electrification in the United States," *The Annals of the American Academy of Political and Social Science* 201 (January 1939):84.

11. *Harvests and Highlines*, 58–9.

12. Ibid., 114–19. *Harvests and Highlines*, published by Insull's Middle West Utilities Company, has no author credit. My assumption is that, since the ghostwriter was in the pay of the company, it accurately reflects the views of the Insulls.

13. Ibid., 67.

14. Gifford Pinchot, *The Power Monopoly: Its Make-up and Its Menace* (Milford, Penn., 1928): 1. This interesting book—a copy of which is in the Huntington Library at San Marino, Calif.—probably was privately printed and distributed. No publisher is mentioned, but Pinchot's name surely guaranteed a wide audience.

15. David E. Lilienthal, "Electricity: The People's Business," *The Annals of the American Academy of Political and Social Science* 201 (January 1939):60.

16. *Harvests and Highlines* 14–5. This figure of 300,000 individual farm plants was considered to be a conservative estimate by the Middle West Utilities Company. The actual number was probably considerably higher, say 500,000.

17. Clara Jensen expressed her dissatisfaction with her windcharger to the author in 1982. She has since passed away.

18. Katherine Kay Jellison, *Entitled To Power: Farm Women and Technology, 1913–1963* (Chapel Hill: Univ. of North Carolina Press, 1993): 184.

19. Quoted in Clyde T. Ellis, *A Giant Step* (New York: Random House, 1966):38–9. Also see the sympathetic biography of Cooke by Kenneth E. Trombley, *The Life and Times of a Happy Liberal: A Biography of Morris Llewellyn Cooke* (New York: Harper & Brothers, 1954):146. For details about the Holding Company Act of 1935, see *Electric Power and Government Policy: A Survey of the Relationship Between the Government and the Electric Power Industry.* (New York: Twentieth Century Fund, 1948):280–4. Also see Richard Lowitt, *George W. Norris: The Persistence of a Progressive, 1913–1933* (Urbana: Univ. of Illinois Press, 1978):83–4, 131–2.

20. Cooke had not only headed the Giant Power Survey, he had also served FDR in New York as a member of the Power Authority of the State of New York. See Platt, *Electric City*, 265, 366; Ellis, *Giant Step*, 39–40.

21. Ibid., 39–40.

22. Trombley, *Life and Times*, 148–9.

23. Ibid., 150–1.

24. See Brown, *Electricity For Rural America*, 67–75.

25. House Committee on Interstate & Foreign Commerce, *A Bill to Provide For Rural Electrification and For Other Purposes*, 74th Cong., 2nd Sess., March 12–14, 1936: 73.

26. L. J. Fletcher, "Rural Electric Service from the Western Standpoint," *Agricultural Engineering* 7 (December, 1926):407. It should be remembered that when Fletcher wrote this piece, the more powerful wind plants, such as the Jacobs or the Wincharger, were not available.

27. David E. Nye, *Electrifying America: Social Meanings of a New Technology, 1880–1940*, (Cambridge, Mass.: MIT Press, 1990): 329.

28. For a wonderfully descriptive account of miserable farm life in the hill country of Texas without electricity, see Robert A. Caro's biography, *The Years of Lyndon Johnson: The Path To Power* (New York: Knopf, 1982):502-515.

29. Nye, *Electrifying America*, 334.

30. Ibid., 328.

31. From interview during visit by author, 1980.

32. Interview with Fred Bruns, June 5, 1987. Bruns was insistent that plenty of pressure was applied by REA cooperative officials, but he could not recall specific incidents.

33. Interview with Paul Jacobs, June 6, 1987. The quote refers to the more recent opposition of utility companies to the PURPA Act of 1978—specifically to their reluctance to allow wind plants to feed electricity into the power systems.

34. See Paul Gipe, *Wind Energy: How to Use It* (Harrisburg, Penn.: Stackpole, 1983): 236.

35. Ellis, *Giant Step*, 57; also see Marquis Childs, *The Farmer Takes a Hand* (New York: Doubleday, 1952). Both are uncritical accounts of the success of REA.

36. F. C. Fenton and D. E. Wiant, "Rural Electrification Surveys of Harvey and Dickinson Counties," *Kansas State College Bulletin* 24 (May 1, 1940):47 pages. Wind-driven plants in the 1000W to 2500W range were Wincharger, Delco, Wards, Sears, and Jacobs.

37. Ibid.

38. In Dorothy Scarborough's novel *The Wind*, for example, the environment (in West Texas) has ravaging effects on a young, Southern-born woman. The wind is a primary source of her anguish and finally breaks her spirit altogether. On the other hand, ownership of a wind machine can alter mood and attitude. Gordon Brittan, rancher, professor, and owner of a 65kW wind turbine, told me it always lifts his spirits to look out his window and see the rotor turning.

39. Nye, *Electrifying America*, 142–3.

40. Don S. Kirschner, *City and Country: Rural Responses to Urbanization in the 1920s* (Westport, Conn.: Greenwood, 1970):247–50.

41. Ibid., 6, 38–9, 41–3.

42. See App, "Industrialization of Agriculture," 233–4. App discusses the inequities of most private utility companies rate schedules when applied to the farmer.

43. Lilienthal, "Electricity: The People's Business," 61–3.

44. House of Representatives, 79th Congress, 1st Sess., *Hearing on Rural Electrification Planning Before a Subcommittee of the Committee on Interstate and Foreign Commerce*, H. R. 1742, June 13, 14, 15; October 16, 17, 18, 19, 30; November 2, 1945: 107–9.

45. Ibid., 109–110.

46. Ibid., 110–3, 115.

47. Ibid., 114–123.

48. Fred W. Hawthorn, "Farm Experience with Wind Electric Plants," *Agricultural Engineering* 19 (January, 1938):7.

49. Sears Roebuck & Company catalogue, spring-fall, 1946. See pages 645 and 1426. Catalogues through 1951 that were examined contained no listings for wind-energy plants.

50. *The Electric Power Industry*, 151. The statistics, provided by the Edison Electric Institute, the successor to the National Electric Light Association, were put together by *Electrical World*.

51. James J. Fenlon, "Current Beyond the Highline," *Successful Farming* 45 (July, 1947):27, 35–6.

52. R. Fincham, "Use Your Windmill For Television," *Successful Farming* 48 (April, 1950):111. Of course, many farmers were also abandoning water-pumping windmills: the article made no distinctions between towers.

53. Jon Naar, *The New Wind Power* (New York: Penguin, 1982):70–71.

54. Interview with Joe and Mary Spinhirne by the author, November 6, 1985.

CHAPTER 6. IN THE DOLDRUMS, BUT PUTNAM AND THOMAS EMERGE

1. "Windmill Electric Power," *Electrical Times* 87 (London, January, 1935): 103.

2. H. E. Parsons, "Wind Power: History and Present Status," *Engineering Journal* 36 (January, 1953):19–21. Writing during the height of cold war rhetoric, Parsons stated that "while loath to credit the Russians with anything they have not already claimed, it would appear that they were the first to tie a wind-power unit into an existing network."

3. Ibid.

4. George A. Whetstone, "What Can Wind Power Do for Us?" *Power Engineering* 55 (March, 1951):72–3

5. Palmer Cosslett Putnam, *Putnam's Power from the Wind*, revised by Gerald W. Koeppl (New York: Van Nostrand Reinhold 1948; 2nd. edn. 1982): 3.

6. Ibid., 4.

7. "World's Largest Wind-Turbine Plant Nears Completion," *Power* 85 (June, 1941):59. Putnam enjoyed a long and close association with Bush, who became one of the illustrious scientists of his age, heading the Office of Scientific Research and Development during World War II. In his autobiography, *Pieces of the Action* (New York: William Morrow, 1970), Bush referred often to Putnam's wartime engineering ability (as noted later in the chapter, Putnam designed landing craft and a snow machine called the Weasel). In admiring Putnam, Bush refers to him as a "go-getter" and an engineer who "well liked by men with lots of drive, and often disliked by those with less." Twice, Bush suggested that Putnam should write an autobiography—it "would be an epic." Unfortunately, that did not happen. See *Pieces of the Action* 104–6, 119, 126–7.

8. John B. Wilbur, "The Smith-Putnam Wind Turbine Project," *Boston Society of Civil Engineers Journal* 29 (July, 1942): 217. Wilbur's work is perhaps the most useful and comprehensive account of the building and the operation of the SmithPutnam plant.

9. Vannevar Bush, "Introduction," in Putnam, *Power From the Wind*, 1948 edition, xi.

10. Wilbur, "Smith-Putnam Wind Turbine Project," 220.

11. "Harnessing the Wind," *Oil-Power* 17 (April, 1942): np.

12. Putnam, *Power from the Wind*, 13.

13. "Harnessing the Wind."

14. Ibid.; also "World's Largest," 57.

15. Palmer C. Putnam, "Wind-Turbine Power Plant Will Be Rebuilt," *Power* 89 (June, 1945):68.

16. The snow machine was similar in design to today's snowmobiles. See Bush, *Pieces of the Action*, for more detail about Putnam's engineering contributions to the war effort.

17. Putnam, "Wind Turbine Will Be Rebuilt," 67.

18. Putnam, *Putnam's Power From the Wind, (2nd edn.)*, 15.

19. Putnam, "Wind Turbine Will Be Rebuilt," 68.

20. Bush, "Introduction," *Power From the Wind*, 1948 edn., xi.

21. Ibid.

22. Wilbur, "Smith-Putnam Wind Turbine Project," 212. Wilbur either was unaware of or ignored the Russian wind generator that pioneered the synchronous converter.

23. Wilbur, "Smith-Putnam Wind Turbine Project," 212.

24. The $300,000 figure comes from a statement by the Bureau of Reclamation in document U.S., Cong., House of Representatives, Committee on Interior and Insular Affairs, *Hearing of H.R. 4286, A Bill. . .to Determine and Demonstrate the Economic Feasibility of Producing Electric Power and Energy By Means of a Wind-Driven Generator*," September 19, 1951: 7.

25. Testimony of William L. Hughes, chair, School of Electrical Engineering, Oklahoma State University, in U.S., Congress, House, "Hearings Before the Subcommittee on Energy of the Committee on Science and Astronautics," 93rd. Cong., 2nd. Sess., May 21, 1974: 60.

26. Hearings on H.R. 4286, 34.

27. Ibid., 32–3.

28. Percy H. Thomas, U.S. Federal Power Commission, "The Wind Power Aerogenerator, Twin-Wheel Type" (Washington, 1946):5, exhibit IV.

29. Thomas testimony on H.R. 4286, 29.

30. Hearing on H.R. 4286, 26–7; Percy H. Thomas, U.S Federal Power Commission, "Electric Power From the Wind" (Washington, DC: 1946) 5.

31. Hearings on H.R. 4286, 27.

32. Ibid., 2–3. 5.

33. Ibid., 16. Although the United States made no effort to substitute wind energy for petroleum energy production during World War II, both the British and the Danes did. The Danes, their country occupied by Nazi Germany, often were without centrally-produced power so they built local wind-electric turbines.

34. Ibid., 5.

35. Ibid., 18.

36. Ibid., 19, 23.

37. Ibid., 7. Later, in the 1970s, the Bureau of Reclamation chose a site (Medicine Bow) reasonable close to Sherman Hills to construct a Boeing MOD-2 2.5MW unit and a Hamilton-Standard 4.0 MW plant. The idea of working in harmony with hydroelectric power remained the same. Perhaps the proponents of H.R. 4286 would have gained political advantage had they picked a site, thus enrolling local and state officials to the cause with the promise of jobs and economic revenue.

38. Ibid., 28–9, 34.

39. Ibid., 21.

40. Quoted in David E. Lilienthal, *Change, Hope, and the Bomb* (Princeton, N.J.: Princeton Univ. Press, 1963):92.

41. Ibid.

42. Ibid., 97.

43. Ibid., 96.

44. Sheldon Novick, *The Electric War: The Fight Over Nuclear Power* (San Francisco: Sierra Club, 1976):38.

45. Lilienthal. *Change, Hope, and the Bomb*, 102–4.

CHAPTER 7. IN THE LEE OF THE ENERGY CRISIS, WIND PICKS UP

1. The rationale of lighting that always remained on was that the unit would far outlast a light that was turned off and on at intervals. Thus, increased use of electricity would be more than offset by lower maintenance costs and fewer unit replacements—at least that was the argument presented when the author questioned the use of twenty-four-hour lighting at the California State Polytechnic University (Pomona) library in 1971. This practice is now definitely on its way out. Consumer advocate Ralph Nader testified before Congress in 1973 on the "gross waste and overillumination" in the use of electricity. Giving examples, he particularly notified that a large building at 1900 Pennsylvania Avenue in Washington, D.C. "has no wall light switches" and that the lights were on twenty-

four hours a day. U.S. Congress, Senate, "Hearings Before the Committee on Commerce . . . on S. 357," 93rd Cong., 1st Sess: 33.

2. Richard Munson, *The Power Makers* (Emmaus, Penn.: Rodale, 1985):105.

3. *Hard energy path* is the antithesis of the low-impact, renewable, clean-energy conservation policies advocated by Amory Lovins in his influential book *Soft Energy Paths: Toward a Durable Peace* (Cambridge, Mass.: Ballinger, 1977).

4. D. G. Shepherd, "Evolution of the Modern Wind Turbine," 47. Typescript held by the author.

5. Jean-Claude Debeir, Jean-Paul Deleage, and Daniel Hemery, *In the Servitude of Power: Energy and Civilization Through the Ages*, translated by John Barzman. (New Jersey: Zed Books, 1991 edn.): 170.

6. Ibid., figures from page 186.

7. Daniel Ford in his critical book, *The Cult of the Atom* (New York: Simon and Schuster, 1982) gives ample evidence gathered from classified AEC materials that in its desire to promote nuclear energy the commission suppressed reports and generally minimized the dangers.

8. John L. Campbell, *Collapse of an Industry: Nuclear Power and the Contradictions of U.S. Policy* (Ithaca: Cornell Univ. Press, 1988): 3–4; Bernard Cohen, *The Nuclear Energy Option: An Alternative for the 90s* (New York: Pleunum, 1990): 145. Cohen has been one of the most articulate proponents of nuclear energy. His views on alternative energy, such as wind, are discussed in note 12. Andrew Blowers and David Pepper, eds., *Nuclear Power In Crisis* (London: Croom Helm, 1987) focuses on the nuclear situation in England.

9. Munson, *Power Makers*, 131.

10. Perhaps the only nation where nuclear power is thriving is France. There, some 70 percent of electrical energy is supplied through nuclear reactors. For a brief analysis and critique of the French system, see Miriam Boyle and Mike Robinson, "Nuclear Energy in France: A Foretaste of the Future," in Blowers and Pepper, *Nuclear Power In Crisis*, 55–84.

11. Ralph Cavanagh, "Recycling Our Electric Utilities," *Amicus Journal* 6 (Summer 1984):28–33. Quote from page 29.

12. Cohen, *Nuclear Option*. In his discussion of wind energy (262–5), Cohen notes that in California in 1987 some 16,000 wind turbines were in use with a capacity of 1,440,000 kilowatts of power—more than that of a large nuclear plant. However, this figure being based on maximum wind speed, Cohen therefore takes the true capacity to be about 25 percent of the kilowatt total. Given that figure, he notes that it would take 50,000 wind turbines to replace the energy produced by a single nuclear or coal-fired plant—"hardly an inviting prospect for an electrical utility" since the "management costs would be horrendous." Cohen notes the efforts to develop larger machines, such as the 3.2MW plant at Kahuku, Hawaii. Whatever progress is made, however, Cohen believes that a wind turbine will never be more than "a fuel-saving device."

13. The environmental movement is a complex topic and beyond the scope of this book. Perhaps the best overview is Samuel Hays, *Beauty, Health, and Permanence: Environmental Politics in the United States, 1955–85* (New York:

Cambridge Univ. Press, 1987). Also see Kirkpatrick Sale, *The Green Revolution*: The American Environmental Movement, 1962–1992 (new York: Hill & Wang, 1993).

14. Gallup International, *The Gallup Opinion Index*, report no. 60 (June 1970): 8, as quoted in Charles O. Jones, *Clean Air: The Policies and Politics of Pollution Control* (Pittsburgh: Univ. of Pittsburgh Press, 1975):153.

15. Quoted in the *New York Times*, February 13, 1971, p. 27.

16. U.S., Congress, Senate, "Hearings Before the Committee on Commerce . . . on S. 357," 93rd Cong., 1st Sess: 1.

17. Excellent works on the oil crisis of the 1970s are Anthony Sampson, *The Seven Sisters: The Great Oil Companies and the World They Made*. (New York: Viking, 1975), John M. Blair, *The Control of Oil* (New York: Pantheon, 1976), Raymond Vernon, ed., *The Oil Crisis* (New York: W. W. Norton, 1976), and Robert Stobaugh and Daniel Yergin, eds., *Energy Future* (New York: Ballantine, 1979).

18. This dramatic story in all its complexity is well told in Sampson, *Seven Sisters*, 230–259.

19. Stobaugh and Yergin, *Energy Future*, 27.

20. "The U.S. Energy Crisis: Some Proposed Gentle Solutions," a paper presented before a joint meeting of the local sections of the American Society of Mechanical Engineers and the Institute of Electrical and Electronics Engineers, January 12, 1972, West Springfield, Mass. Reprinted in the *Congressional Record* V, 118, pt. 3, February 7, 1972: 3587–3592.

21. Ibid., 3588. Heronemus's barbs were aimed at Chauncey Starr's article in the *Scientific American* (September 1971). This article, "Energy and Power," was reproduced in *Energy: Readings From Scientific American*, S. Fred Singer, ed. (San Francisco: W. H. Freeman, 1979). Starr minimized the role that wind power could play in the future energy mix of the United States.

22. Ibid. Also see Volta Torrey, *Wind Catchers: American Windmills of Yesterday and Tomorrow* (Brattleboro, Vt.: Stephen Greene, 1976): 166–9, for the influence of Heronemus on wind energy.

23. U. S., Congress, Senate, *Hearings Before the Committee on Interior and Insular Affairs Pursuant to S. Res. 45, A National Fuels and Energy Policy Study*, 93rd Cong., 1st Sess: 11. President Nixon's energy message, in its entirety, is on pages 3 to 20.

24. Ibid., 17, 607–8. Perhaps Nixon's $4 million baseline was monies being spent on some sort of atmosphere research. It certainly was not going into wind-electric turbines.

25. See U.S. Senate, Committee on Interior and Insular Affairs, *Hearings on S. 3234, Solar Energy Research Act of 1974*, 93rd Cong., 2nd Sess., June 27, 1974: 3–14, passim. SERI was established under Public Law 93-473.

26. To give an example of the increase in funding, ERDA spent an estimated $15.1 million in 1976 on wind energy. In 1977, the agency was appropriated $21.4 million. See U.S. Senate, Committee on Interior and Insular Affairs, *Congress and the Nation's Environment: Energy and Natural Resources Actions of the 94th*

Congress, prepared by the Environment and Natural Resources Policy Division, Congressional Research Service, Library of Congress, 95 Cong., 1st Sess., January 1977: 357–8. The $380 million figure is taken from a paper delivered at a conference of the Society for the History of Technology in Cleveland, October 18–21, 1990, by Matthias Heymann, "Why Were the Danes Best? Wind Turbines in Denmark, West Germany and the USA, 1945–1985": 1, fig. 1.

27. Quoted in "Power in the Wind," *Newsweek* (February, 1978), 78.

28. Ibid., 103.

29. Frederick March, et al., *Wind Power for the Electric-Utility Industry* (Lexington, Mass.: D.C. Heath, 1982): 9–12.

30. March, *Wind Power*, 12.

31. S. J. Hightower, "The Bureau of Reclamation's Wind Program," in *Wind Energy in Rural America*, proceedings of the Rural Electric Wind Energy Workshop, June 1–3, 1982, Boulder Colorado: 132.

32. William R. Johnson, "Design, Construction, and Early Operation of the 3.2-MW MOD-5B Wind Turbine," in *Windpower 87'*, Proceedings of the American Wind Energy Association, San Francisco, October 5–8, 1987: 1-6.

33. See U.S. House of Representatives, Subcommittee on Energy Development and Applications of the Committee on Science and Technology, *Hearings on the Wind Energy Systems Act of 1980*, 96th Cong., 1st Sess., September 18, 24, 26, and October 17, 1979: 298–9. This hearing is particularly helpful in understanding how Congress and the large companies viewed the national wind program. Officials from DOE testified, as did representatives from Boeing, General Electric, Hamilton Standard, Grumman, and others, such as the smaller company of Windworks, Inc.

Although U.S. engineers boasted that the MOD 5B represented the largest generator in the world, in fact, Germany's Growian wind turbine, with a rotor that was one hundred meters in diameter, was larger.

34. U.S. Congress. House of Representatives, Subcommittee on Energy Development and Applications of the Committee on Science and Technology, *Hearings on Oversight Wind Energy Program*, 96th Cong., 1st Sess., July 30, 1979: 98–102.

35. See Barry M. Casper and Paul David Wellstone, *Powerline: The First Battle of America's Energy War* (Amherst: Univ. of Massachusetts Press, 1981) for the remarkable story of a group of committed farmers who fought, both in words and deeds, the REA bureaucracy's dedication to huge central power schemes.

36. E. F. Schumacher, *Small Is Beautiful: Economics As If People Mattered* (New York: Harper & Row, 1973); also influential was Amory Lovins, *Soft Energy Paths*.

37. Story in the *Denver Post*, July 15, 1973, 34.

38. I make an apology to all those wind-energy pioneers who cannot be mentioned in this work— windsmiths who have devoted much of their lives to proselytizing for use of the "benevolent breeze."

39. Letter from Paul Gipe to the author, August 8, 1993.

40. Interview with Paul Gipe by the author, November 11, 1992.

41. See Paul Gipe, *Wind Energy: How to Use It* (Harrisburg, Penn.: Stackpole, 1983): 234–241, for accounts of his Montana wind-turbine hunts.

42. Interview with Paul Gipe by the author, November 11, 1992.

43. Gipe, *Wind Energy.*

44. Interview with Ty Cashman by the author, November 16, 1992.

45. Ibid.

46. Ibid. Jerry Brown has taken little credit for the development of wind energy development in California. When he ran for the Democratic presidential nomination in 1992, he hardly mentioned his effort in renewable energy, prompting writer Peter Asmus to ask, "Why Doesn't Jerry Brown Tout California's Success Story?" See the opinion page of the San Diego *Union-Tribune*, March 11, 1992.

47. Interview with Fred Bruns by the author, June 5, 1987.

48. *Minnesota State Times*, April 12, 1984, 5B.

49. Ibid., February 7, 1985, 6B. Also see letter from Steve Turek, Wind Turbine Industries, to the author February 4, 1993. In a letter from Paul Jacobs, Marcellus's son and president of Jacobs Wind Electric Co., to the authror, May 10, 1993, Paul Jacobs said he and a number of engineers were "recreating a new firm on the foundation of our Florida firm."

50. Interview with Ty Cashman by the author, November 16, 1992.

51. Roberta Walburn, "Breezy Old Pro is Back," *Minneapolis Tribune*, May 25, 1980, 1B.

52. U.S. Congress, House of Representatives, *Wind Energy Hearing Before the Subcommittee on Energy of the Committee on Science and Astronautics*, 93rd Cong., 2nd Sess., May 21, 1974; 207.

53. M. L. Jacobs, Jacobs Wind Electric Co., to Rep. John R. Murdock, chairman, Committee on Interior and Insular Affairs, October 5, 1951, in U.S. Congress. House of Representatives, *Hearings on H.R. 4286, "Production of Power by Means of Wind-Driven Generator,"* 82, Cong., 1st Sess., September 19. 1951: 40–1.

54. *Tri-Valley Herald* (Livermore, Calif.), February 14, 1983, 1

55. Walburn, "Breezy Old Pro," 4B.

56. Quote from interview with Ty Cashman by the author, November 16, 1992.

CHAPTER 8. THE RIDDLE OF RELIABILITY

1. Andrew Trenka, SERI, DOE, as quoted in the *Boulder Daily Camera*, Colorado, September 2, 1984, 8A.

2. Bill Adams of San Gorgonio Farms, as quoted in the *Desert Sun,* Palm Springs, March 30, 1990, A1, A13.

3. For an exposé on Dynergy see Anne Richards, "Wind Firm Short of Cash, Power, long on Lawsuits," Palm Springs *Press-Enterprise*, May 18, 1986, B1-4: Also see the *Desert Sun*, Palm Springs, January 9, 1988, 1.

4. Report in *Alternative Sources of Energy* 50 (July-August, 1981):38.

5. Ibid.

6. Richard O'Reilly, "Wind Energy Plans Weather the Storms," *Los Angeles Times*, February 21, 1982, sec. II, 1.

7. Donald Marier, "Overview: The Developing Wind Industry," *Alternative Sources of Energy* 58 (November-December, 1982):8–14. According to Marier, Marvin Cheney of UTRC decided to form his own company, using the UTRC design (now called Windtech).

8. Carol DeWindel, "WPL's Wind Energy Test Program: The First Year," *Alternative Sources of Energy* 58 (November-December, 1982):16–19.

9. Bill Stall, "Wind Power Researched," *Denver Post*, August 24, 1977.

10. Terry J. Healy and Darrell M. Dodge, "Characteristics of Small Wind Systems," in *Wind Energy in Rural America*, proceedings of the Rural Electric Wind Energy Workshop, June 1–3, 1982, Boulder, Colo., sponsored by DOE Wind Energy Tech.SERI: 137–49.

11. Ibid., 146.

12. Mogens I. Rasmussen as quoted in Michael Henzl, "New Mexico Wind Energy Czar," *New Mexico* 60 (February, 1982):44--48. Quoted from page 44.

13. James Schmidt, "Wind System Experiences: Overcoming the Technology Lag," *Alternative Sources of Energy* 50 (July-August, 1981):10–12.

14. Todd Malmsbury, "Turbines Spinning an Ill Wind," *Boulder Daily Camera*, Colorado, September 2, 1984, 1A, 8A.

15. During the early 1980s, when I was asked by interested persons if they should purchase a wind generator, my answer was no, not unless they had plenty of capital, a good knowledge of electricity, and enjoyed tinkering with machinery.

16. Barry N. Haack, *An Examination of Small Wind Electric Systems in Michigan* (Ann Arbor: Univ. of Michigan, Geographical Publication no. 21, 1977). Haack based his comments primarily on cost, but reliability was also a factor.

17. U.S. Congress, House of Representatives, *Hearing Before the Subcommittee on Energy Development and Applications of the Committee on Science and Technology, Oversight, Wind Energy Program*, 96th Cong., 1st. Sess., July 30, 1979: 155, 163, 178–186, passim.

18. G. P. Tennyson, "Potential of Wind Energy Conversion Systems in the Great Plains," in *Proceedings of the Solar and Wind Systems Workshop* (Lincoln, Neb.: Great Plains Agricultural Council Energy Committee, 1983):78.

19. *New York Times*, Sunday, June 27, 1976.

20. Richard O'Reilly, "Wind Energy Plans Weather the Storms," *Los Angeles Times*, February 21, 1982, sec. II, 1.

21. Zeke Scher, "Putting Wyoming Wind to Work," *Empire Magazine* (Sunday supp., *Denver Post*), July 15, 1979, 10, 14.

22. "Wind Power from Medicine Bow," 13-page pamphlet published by the Bureau of Reclamation: 11.

23. This account is taken primarily from S. J. Hightower, "The Bureau of Reclamation's Wind Program," in *Wind Energy In Rural America*: 129–134.

24. Figures from "Wind Power from Medicine Bow": 10.

25. *Laramie Daily Boomerang*, May 1, 1985, 15; *Laramie Daily Boomerang*, September 27, 1986, 1.

26. Kit Miniclier, "Nobody Wants 350-foot Windmill Because of $1.5 Million Repair Bill," *Denver Post*, May 30, 1985; also see *Laramie Daily Boomerang*, May 1, 1985.

27. Description based on Toby F. Marlatt, "Malfunction Shatters World's Largest Wind Turbine," *Medicine Bow*, Wyoming, January 20, 1994, 1, and *Laramie Daily Boomerang*, Sunday, January 16, 1994.

28. Gerald W. Braun, Don R. Smith, "Commercial Wind Power: Recent Experience in the United States," *Annual Review Energy Environment* 17 (1992): 97–121, esp. page 105.

29. The overall wind program was cut from $80 million in FY81 to $36 million in FY82. Figures from Tennyson, "Potential":78.

30. *Laramie Daily Boomerang*, September 27, 1986, 1; Conversation with David Spera by author, October, 1985.

31. Most experts and wind-energy company employees interviewed by the author felt that the MOD series, (both first and second generation) were impractical in design and would never succeed in a commercial setting.

32. See "Planning Outline," in U.S. House of Representatives, Subcommittee on Energy Development and Applications of the Committee on Science and Technology, *Hearings on the Wind Energy Systems Act of 1980*, 96th Cong., 1st Sess., September 18, 24, 26, and October 17, 1979: 286–315.

33. Carol Dewinkel, "WPL's Wind Energy Test Program: The First Year," *Alternative Sources of Energy* 58 (November-December, 1982):19.

34. I am in debt to Matthias Heymann of the Deutsches Museum, Munich, for correspondence and conversation. Most of the data and ideas expressed in this paragraph may be found in Heymann's papers: "Why Were the Danes Best: Wind Turbines in Denmark, West Germany and the USA, 1945–1985," and "Theoretical Knowledge and Experimental Skill: What Made Wind Energy Converter Successful in History and at Present?"

35. Heymann, "Why Were the Danes Best," figs. 3, 5.

36. Ibid., fig. 10.

37. Interview with Gipe by author, November 11, 1992.

38. Comparison provided by Andrew Swift, mechanical engineer, Univ. of Texas at El Paso.

39. Trade literature provided by R. Lynette and Associates and Advanced Wind Turbines, Inc.

CHAPTER 9. DECISIONS FOR THE WIND

1. Article on the wind turbine in the *New York Times*, November 13, 1976, 27.

2. This account is taken primarily from *Windmill Power For City People: A Documentation of the First Urban Wind Energy System* (Energy Task Force, New York City, May 1977). The booklet not only has a strong social message in its sixty-five pages; it also packs in information on windmill engineering, urban wind patterns, installation, urban building codes, and simple ways to economize.

3. *New York Times*, November 13, 1976, 27.

4. *Windmill Power For City People*, 34.

5. "Linda Greenhouse, "State Tells Con Ed to Buy 2 Kilowatts—From a Windmill," *New York Times*, May 6, 1977, A1, D13.

6. Ibid.

7. *New York Times*, May 8, 1977, E6.

8. *Windmill Power for City People*, 43.

9. Ibid., Foreword.

10. *Congressional Record*, Vol. 123, Pt. 18, 95th Cong., 1st Sess: 22297.

11. Richard Munson, *The Power Makers* (Emmaus, Penn.: Rodale, 1985): 118–20.

12. Quoted in Martin Merzer, "Windmill Power Up In Air," *Denver Post*, June 15, 1977.

13. Clipping of letter to the editor from Karen Pamperin and article clipping, John Nelander, "Windmill Stirs Utility Bill Battle," in Jacobs Wind Energy Co. files, folder 106, Panhandle-Plains Historical Museum, Canyon, Texas. These clippings were from an unidentified Wisconsin newspaper of 1978 (possibly 1979). In the earlier 1970s, opposition to hookups of independent power were much more pronounced. The Nelander clipping mentioned that the Northern States Power Company, based in Eau Claire, Wisconsin, had no problem allowing the meter to run backwards, thus accommodating independent power producers.

14. See Merzer, "Windmill Power Up In Air."

15. *New York Times*, September 1, 1964, 37.

16. "California and the Human Horizon," in *The Urban Prospect*, essays by Lewis Mumford (New York: Harcourt, Brace & World, 1968);9.

17. "California Energy Futures: Two Alternative Societal Scenarios and Their Energy Implications," California Energy Commission, P300-80-027, June 1980, S-6.

18. Richard F. Hirsh, *Technology and Transformation in the American Utility Industry* (New York: Cambridge Univ. Press, 1989):155.

19. State of California. *Energy Dilemma: California's 20-Year Power Plant Siting Plan* (Sacramento, June 1973):107.

20. *Energy Alternatives For California: Paths to the Future*. Rand Corporation report prepared for the California State Assembly, December 1975:iii. In 321 pages.

21. Ibid., 145, 149.

22. Ibid., 236. The ten potential new energy sources for California were (1) offshore oil, (2) oil from Alaska, (3) natural gas from Alaska, (4) liquid natural gas (LNG) from Indonesia and other sources, (5) nuclear, (6) geothermal, (7) coal-fired plants in the Southwest, (8) solar (but not including wind), (9) organic materials, (10) natural gas from coal.

23. See David Roe, *Dynamos and Virgins* (New York: Random House, 1984):27.

24. This story is told with great detail by David Roe in *Dynamos and Virgins*.

25. Ibid., 71. Also see page 120.

26. Ibid., 101.

27. Ibid., 120–2.

28. Ibid., 196–7.

29. Rick Gore, "Conservation: Can We Live Better On Less?" in "Special Report on Energy", *National Geographic* (February 1981):34–57. Quote from page 44.

30. Roe, *Dynamos and Virgins*, 102.

31. "The California Covenant," *Windpower Monthly* 5 (July 1989):18–20.

32. Robert Thomas, president, Wind Harvest Company, letter to the author, February 10, 1993.

33. Account largely based on an interview with Ty Cashman by the author, November 16, 1992.

34. No relation to Gary Hart of Colorado.

35. California, *Statutes and Amendments to the Codes*, 1978, Volume 3, Chapter 1159: 3557.

36. Interview with Ty Cashman by the author, November 16, 1992.

37. Munson, *Power Makers*, 37.

38. U.S. Code. Congressional and Administrative News, 95th Cong., 2nd Sess., 1978, *Laws*, Vol. 2, Public Law 95-617, Sec. 210.

39. Quote from a congressional staff member in Munson, *Power Makers*, 38.

40. U.S. Code. Congressional and Administrative News, 95th Cong., 2nd Sess., 1978, *Legislative History*, Vol. 6:7831–3.

41. Ibid., 7832.

42. U.S. Code. Congressional and Administrative News, 95th Cong., 2nd Sess., 1978, *Laws*, Vol. 2, Public Law 95-617, Sec. 210, (d) Definition.

43. See Frederic March, et al., *Wind Power for the Electric-Utility Industry* (Lexington, Mass.: D. C. Heath, 1982): 77–82, for a detailed discussion of the PURPA law and the concept of avoided costs.

44. March, *Wind Power*, 79.

45. Ralph Cavanagh, "Recycling Our Electric Utilities," *Amicus Journal* 6 (summer, 1984):33.

46. Melinda Welsh, "Battle Lines Being Drawn Up," *Windpower Monthly* (August 1986): 13 and 26.

47. Melinda Welsh, "Utilities Threaten US Wind Energy Business," *Windpower Monthly* (August 1986): 12–13). Quote from page 12.

CHAPTER 10. CALIFORNIA TAKES THE LEAD

1. *Congressional Record*, V, 126, Pt. 18, 96th Cong., 2nd Sess: 23110.

2. Ibid., 23110

3. Richard F. Hirsh, *Technology and Transformation in the American Electric Utility Industry* (New York: Cambridge Univ. Press, 1989):181–2.

4. Ibid.

5. *Wind Energy Program Progress Report*, January 1, 1980 (California Energy Commission (CEC), 500-80-001): 1. One plan was drawn up in 1979 by a Ph.D. student at the University of California, Los Angeles. The plan would locate five thousand 2MW wind turbines on one hundred sites, providing 10,000MW of rated wind-electric capacity. See Matania Ginosar, "A Large Scale Wind Energy Program for the State of California" (UCLA, dissertation in

Environmental Science and Engineering, 1979). It turned out that the 1987 target was not met, but by 1991 the state was closing in on the 1 percent figure.

6. *Wind-Electric Power: A Renewable Energy Resource For California* (CEC, 500-78-025, July, 1978):14.

7. "The California Covenant," *Windpower Monthly* 5 (July, 1989):20.

8. Ibid. Twenty states offered tax credits for wind-energy development. Most of these states limited credits to SWEC systems—usually a limit of up to $5,000. However, six states, including Arkansas, Hawaii, North Dakota, Ohio, and Oklahoma, had no tax-credit limit, opening the way for commercial development.

9. Christopher Flavin, "Wind Energy: A Turning Point," *Worldwatch Paper 45*, July, 1981: 38.

10. Solar Energy Conversion Systems, Inc. *Impact of Large Wind Energy System In California* (Glendora, Calif.: CEC, Contract 500-1-11(7/8), June 1980): 3-1, 3–21.

11. U.S. Congress, House of Representatives, *Hearings Before the Subcommittee on Energy Development and Application of the Committee on Science and Technology, Wind Energy Systems Act of 1980*, 96th Cong., 1st Sess., September 18, 24, 26, and October 17, 1979:75. Two significant studies on the wind potential of California were Albert Miller and Richard L. Simon, *Wind Power Potential in California* (Department of Meteorology, San Jose State University, Contract 500-09(7/8) with the CEC, 1978) and James D. Goodridge and Earl G. Bingham, *Wind In California* (Bulletin 185, California Department of Resources, 1978).

12. Approximately 85 percent of electricity consumed in California is provided by PG&E, SCE, San Diego Gas and Electric, Los Angeles Department of Water and Power, and Sacramento Municipal Utility District.

13. *Impact of Large Wind Energy Systems*, 5–4, 5–5.

14. *Potential Pumped Storage Projects in the Pacific Northwest*, Federal Power Commission, Bureau of Power, 1975. Figures quoted in *Impact of Large Wind Energy Systems*, 5–17, 5–18.

Today, 4MW turbines are practicable, although no one is building machines of that size. A 500 kW is considered reasonable, with 1MW on the horizon.

15. *Impact of Large Wind Energy Systems*, 1.

16. *A Guide To Financial Assistance For Wind Energy*, (CEC, P500-81-014, March, 1981):8–9.

17. Ibid., 35–6.

18. Ibid., 31–2

19. Ibid., 40.

20. *Model Ordinance for Small Wind Energy Conversion Systems*. CEC, Office of Appropriate Technology, April 1982:1, 14–17.

21. CEC, *News and Comments* 4 (July/August 1981):3–4.

22. Robert Teitelman, "Technology," *Forbes* (July 18, 1983):134.

23. "WindMaster Partners 1983-1: A California Limited Partnership," confidential private offering memo, BEPW Development Corporation, June 1983, 6–7. Copy held by the author.

24. "WindMaster Partners 1983-1," supplement to BEPW, July 22, 1983: 4.

25. Author interview with Asa Barnes, November 17, 1992.

26. Ad provided to the author by Robert Kahn & Co.

27. CEC, *News and Comments* 1 (November 1981):3, 14.

28. Michaela Jarvis, "Power From Thin Air," *California Journal* (July 1989): 278–283.

29. Information from "Wind Power Plants in the Tehachapi-Mojave Wind Resource Area," a Kern Wind Energy Association brochure.

30. The American Wind Energy Association *Membership Directory*, 1992, lists International Turbine Research, Inc., as owning and operating 163 turbines at Pacheco Pass. See page 16.

31. *International Market Evaluations: Wind Energy Prospects*, CEC, P500-87-004, 1987: 1–16.

32. "Altamont Pass Wind Power Plant," PG&E brochure, April 1992.

33. Ibid.

34. Based on a Boston *Globe* story reprinted in the Los Angeles *Times*, June 8, 1981, Sec. IV, 2.

35. Ibid.

36. Interview with Cashman by the author, November 16, 1992.

37. *International Market Evaluations: Wind Energy Prospects*. CEC, P500-87-004, 1987: 13.

38. Interview with Gipe by the author, November 11, 1992.

39. Ibid.

40. Figures from "California Wind Industry Development," Paul Gipe & Associates, as compiled from CEC annual reports, March 10, 1994; Also see *International Market Evaluations*, 18.

41. See Ros Davidson, "Environmental Issue In Need Of Reality Not Romanticism," *Windpower Monthly* 3 (October, 1987):16–18.

42. "Gone With The Wind," *Time Magazine* 127 (January 20, 1986):23.

43. Birger Madsen, "Where To Now?" *Windpower Monthly* 4 (November 1988):6.

44. "Death of a Pioneer," *Windpower Monthly* 4 (July, 1988):14.

45. Ros Davidson, Jos Van Beek, and Torgny Moller, "Windfarms Shut Down As Crisis Worsens," *Windpower Monthly* 4 (February 1988):8; Ros Davidson, and Lyn Harrison, "Micon Sued," *Windpower Monthly* 4 (March 1988):8–10.

46. Ros Davidson, "Uneasy and Divisive in the Face of Survival," *Windpower Monthly* 4 (November 1988):14–5.

47. Ibid.

48. Ibid., 17.

49. Ibid., 20. The company has also serviced these turbines. As noted in Chapter 11, they have now been sold to New World Grid Power, and some are still operating.

50. Ibid., 24.

51. The 500kW is the consensus of a survey of engineers by Andrew Swift, Department of Mechanical Engineering, University of Texas at El Paso. The survey also predicts a larger size for Europe because of limitations on land area available.

52. Davidson, "Uneasy," 22. Also see PG&E brochure, "Altamont Pass."

53. Lon House, "Wind Farm Avoided Cost: A California Example," in CEC, *News & Comments* 12 (fall 1983):7, 13.

54. Ibid.

55. Michaela Jarvis, "Power From Thin Air," *California Journal* (July 1989):283.

56. Interview with Gipe by the author, November 11, 1992.

CHAPTER 11. CATCHING THE BREATH OF THE SUN: PROBLEMS AND PROMISE

1. Quoted in Michaela Jarvis, "Power From Thin Air," *California Journal* (July 1989):281.

2. Robert L. Scheffler, "Wind Technology Comes of Age—Part 1," *R&D Newsletter*, SCE, summer, 1986: 6.

3. *Los Angeles Times*, October 24, 1980, sec. 2, 4.

4. SCE, "Research and Development Fact Sheet, Wind Turbine Generation Test Program." Provided by the office of Robert Scheffler, SCE senior research engineer.

5. Steve Moore, "Edison Wind Turbine Begins Producing Power," Palm Springs *Press-Enterprise*, December 17, 1980, n. p. This news story as well as many others on wind energy may be found in the "Wind Energy" clipping file, Palm Springs Public Library.

6. Quoted in the *Desert Sun*, March 30, 1990, A1, 13.

7. Some of the information presented is based on Anne Richards, "Wind Firm Short of Cash, Power, Long on Lawsuits," Palm Springs *Press-Enterprise*, May 18, 1986, B-1,4.

8. *Desert Sun*, January 9, 1988. In wind energy clipping file, Palm Springs Public Library (hereafter, PSPL).

9. In 1990, Judith Golden, a spokeswoman for the IRS, Laguna Niguel, said that the IRS filed dozens of complaints against early wind-farm developers in the San Gorgonio Pass, but most were settled out of court. *Desert Sun*, March 30, 1990, A-1, 13.

10. Richards, "Wind Firm Short of Cash," Palm Springs *Press-Enterprise*, May 18, 1986, B-1, 4.

11. Ibid.

12. *Desert Sun*, July 2, 1987, A-3.

13. Anne Richards, "Ill Winds Over Windmills," *Golden State Report* (February, 1987):35–6.

14. *Desert Sun*, May 19, 1987, A-3.

15. *Desert Sun*, January 9, 1988. In wind energy clippings file, PSPL.

16. Interview with Dave Kelly, field manager of SeaWest Tehachapi, by the author, November 10, 1992.

17. *Desert Sun*, February 23, 1988, A-1.

18. Mayo Mohs, "Blowin' in the Wind," *Discover: The Magazine of Science*, (June 1985):70.

19. "Former Pop Star Fights Windfarm Plans," *Windpower Monthly* 4 (April 1989):16.

20. Letter to the editor from Mrs. Lee E. Griffin, *Desert Sun*, July 8, 1989, E-4.

21. Robert J. Hutchinson, "Transforming the Sea Wind," *Oceans* (September-October, 1987):12–19. Quote from page 16. This article focuses on wind-energy development in Hawaii.

22. *Desert Sun*, June 3, 1989, in wind energy clippings, PSPL; also see *Desert Sun*, July 1, 1989, E-1; and Ros Davidson, "Sony Bono Takes His War on Windmills to Washington," *Windpower Monthly* 5 (July 1989):21.

23. Martin J. Pasqualetti and Edgar Butler, "Public Reaction to Wind Development in California," *International Journal of Ambient Energy* 8 (April 1987):83–89.

24. Ibid., 89.

25. *Desert Sun*, August 1, 1990, A-1.

26. These and other industry leaders formed the association in mid-September 1989. Miles Barrett, coowner and vice president of Wintec, stated the purpose was "to counteract myths and negativism." See Ros Davidson, "Desert Wind Group Fights Back," *Windpower Monthly* 5 (October, 1989):14.

27. *Desert Sun*, January 28, 1989, A-6.

28. Letter to the editor from Clare Lees, *Desert Sun*, June 17, 1989, I:3.

29. *Desert Sun*, June 29, 1989, A-12.

30. "Vonda's Valley" column, *Desert Sun*, July 20, 1991, in wind energy clippings, PSPL.

31. *Desert Sun*, December 5, 1991, A-3.

32. *Desert Sun*, August 1, 1990, A-1.

33. *Los Angeles Times*, April 13, 1991, in wind energy clippings, PSPL.

34. Interview with Curt Maloy by the author, November 9, 1992.

35. *Desert Sun*, June 10, 1989, F-1.

36. Information from a three-part series by Jeffrey Potts, "Tilting at Windmills," *Desert Sun*, March 29, 30, 31, 1990.

37. Ibid.

38. Interview with Maloy, November 9, 1992.

39. Interview with Scheffler by the author, November 18, 1992.

40. Interview with Gipe by the author, November 11, 1992.

41. Information from Zond Systems trade literature.

42. Andrew C. Revkin, "Firm Plans to Use Wind-Powered Turbines to Reap Energy on Range," *Los Angeles Times*, January 11, 1987, W-1, 5.

43. Paul Gipe, "Zond Fights Propaganda With Profit Share Offer," *Windpower Monthly* 5 (January, 1989):6–7.

44. *Los Angeles Times*, January 23, 1989, sec. II, 1.

45. *Bakersfield Californian*, July 30, 1989, in Bakersfield Public Library, windmills clipping file (hereafter BPL clippings).

46. *Bakersfield Californian*, October 24, 1988, BPL clippings.

47. *Bakersfield Californian*, July 30, 1989, BPL clippings.

48. Steve Padilla, "County Planners Reject Wind Farm for Gorman Area," *Los Angeles Times*, August 17, 1989, sec. 1, 3; *Bakersfield Californian*, August 27, 1989, in BPL clippings file.

49. Robert L. Thayer Jr. and Heather A. Hansen, "Wind Farm Siting Conflicts in California: Implications for Energy Policy," 43-page pamphlet published by the Center for Design Research, Univ. of California, Davis, May 1991: 31.

50. Quoted in "Field of Windmills turns Tehachapi into power Mecca," Las Cruses *Sun-News* (New Mexico), March 15, 1993, A-6. Reprint of article by the Los Angeles *Daily News.*

51. Alston Chase, "California Wind Farm Fight Generates Unusual Foes," *Jackson Hole Guide*, August 23, 1989.

52. Quoted in Thayer, Hansen, "Wind Farm Siting," 28.

53. Ibid., 32–3.

54. "California Wind Industry Development," Paul Gipe & Associates, March 10, 1994.

55. Robert L. Thayer Jr., "The Aesthetics of Wind Energy in the United States: Case Studies in Public Perception," *Proceedings*, European Community Wind Energy Conference, Herning Congress Centre, Denmark, June 6–10, 1988 473.

56. Michael Collier, "Altamont Ranchers Cast Their Fate to the Winds," *Tri-Valley Herald* (Livermore), January 30, 1983, 4.

57. D. R. Smith, "The Wind Farms of the Altamont Pass Area," *Annual Review: Energy* 12 (1987): 163.

58. Thayer, "Aesthetics," 167–8. Paul Gipe agreed with criticisms about the ESI: "If you had to pick an ugly windmill to design, you couldn't pick a better one than the ESI." Interview with Gipe, November 11, 1992.

59. Ibid., 164.

60. *Tri-Valley Herald*, July 12, 1991, B-1.

61. Ibid.

62. Sylvia White, "Towers Multiply, and Environment is Gone with the Wind," *Los Angeles Times*, November 26, 1984, sec. II, 5.

63. *Tri-Valley Herald*, November 21, 1984, 12.

64. Ibid. The Teamsters Union position no doubt was influenced when, in September 1983, thirty employees of U.S. Windpower voted against joining the union.

65. *Tri-Valley Herald*, March 22, 1984, 13.

66. Carl Blumstein to Randall M. Ward, executive director, CEC, December 18, 1985, in "Solar and Wind Technology Tax Incentive Impact Analysis, Reviewer's Comments, Appendix A," CEC, P500-86-010A, 1986, 1–2.

67. Interview with Paul Gipe by the author, November 11, 1992. That month, the new federal energy act provided a 1.5 cent per kWh subsidy on *electricity produced*—a more responsible method of subsidizing the production of renewable energy.

68. Seth Zuckerman, "Winds of Change," *Image* (September 20, 1987):30.

69. Ibid., 29.

70. White, "Towers Multiply," sec. II, 5.

71. Ibid., 30.

72. Jeff Greenwald, "Reaping the Wind: The Farmers of Altamont Pass," *Image* (June 8, 1986):30, 32.

73. "Fields of Windmills Turns Tehachapi into power Mecca," Las Cruces *Sun-News* (New Mexico), March 15, 1993, 1-6.

74. Robert L. Thayer and "Technophobia and Topophilia: The Dynamic Meanings of Technology in the Landscape," paper in the possession of the author. Also see Thayer, "Dimensions and Dynamic Meanings of Technology in the American Landscape," *IS Journal* no. 10 (fall 1990):19–38. Thayer's recent book, *Gray World, Green Heart: Technology, Nature, and the Sustainable Landscape* (New York: John Wiley, 1994) focuses on such topics as resource consumption and "landscape guilt," seeking solutions in "sustainable landscapes" that conserve resources and ecological characteristics.

75. Leo Marx, *The Machine in the Garden* (New York: Oxford Univ. Press, 1964). For the types of "landscape guilt" and the ways in which society deals with technophobia, see Thayer, *Gray, Green* 66–79.

76. Robert L. Thayer and Carla M. Freeman, "Altamont: Public Perceptions of a Wind Energy Landscape," *Landscape and Urban Planning* 14 (1987):385.

77. Ibid., 386–96.

78. "Avian Mortality at Large Wind Energy Facilities in California: Identification of a Problem" (California Energy Commission, P700-89-001, August, 1989).

79. Ibid., 18.

80. Quoted in the *Bakersfield Californian*, May 8, 1992, in BPL clippings. In November 1992, on the Old Altamont Pass Road, the author observed at least half a dozen turkey vultures soaring around and through two slow-turning FloWind vertical turbines. The vultures are numerous in the region, yet there are few reported deaths—whereas eagles and hawks are killed. The buzzards seem to have adjusted to the intrusion of turbines. The difference may be attributed to species characteristics: buzzards leisurely circle before claiming their dead prey, whereas hawks and eagles focus on prey and rapidly dive.

81. Interview with Gordon Brittan, president of Montana Wind Turbine, Inc., by the author, October 3, 1992. Much of the story of Livingston's interest in wind energy has been related to the author by Brittan, who is intimately acquainted with this novel community effort.

82. Paul Laird and Ed Stern, "Livingston Wind Energy System: A Municipal Commercialization Project: June 1981–November 1982," (Montana Department of Natural Resources and Conservation, December 1982):6–8.

83. Ibid., 23–6.

84. *Alternative Sources of Energy* 55 (May–June 1982):34.

85. Ibid.

86. Ibid.; also two brochures, "Montana Wind Turbine, Inc." and "The Windjammer 4". The International Harvester Company, which funded the first Windjammer, has reformed as Navistar.

87. See "Montana Fights On," *Windpower Monthly* 4 (July 1988):14.

88. Ed Stern, "Local Governments and the Wind Energy Industry: A New Partnership," in *Proceedings*, Wind Energy Expo '82, American Wind Energy Association, Amarillo, Texas, October 24–27, 1982, 75–77.

89. Livingston *Enterprise*, March 13, 1992, 6.

90. Ibid.

91. Researcher Robert Thayer has found the "residents who have recently moved from urban areas will most strongly disfavor wind development." Thayer, "Aesthetics," 473.

92. Livingston *Enterprise*, April 10, 1992, 1.

93. Ron Wiggins to the author, April 3, 1993.

94. Todd Wilkinson, "Raptors, my friend, are flyin' in the wind," *Jackson Hole News*, March 18, 1992, 4-B.

95. "Governor Carlson Dedicates New Minnesota Wind Project," *Windletter*, American Wind Energy Assoc. no. 6, June 1992: 1, 5.

96. "Minnesota Windpower Has Designs to Power the Midwest," *Windletter* 20 (March, 1993): 5; also see Minneapolis *Star-Tribune*, October 30, 1993, May 31, 1994, June 2, 1994.

97. R. Richard Neill, "Progress Report on Wind Energy Programs in Hawaii, in "Windpower 87," *Proceedings*, annual conference of the American Wind Energy Association, San Francisco, October 5–8, 1987: 281–286.

98. See *Windletter* 1, 1988: 1–2. For details on the specifications and the erection of the MOD 5B see William R. Johnson, "Design, Construction and Early Operation of the 3.2-MW MOD 5B Wind Turbine," Windpower '87, *Proceedings*: 1–6.

99. Neill, "Progress Report," 285–6.

100 *Windletter* 4, 1986, 2–3.

101. *Windletter* 7, 1986: 3–4.

102. Interview with Paul Gipe by the author, November 11, 1992. Both the Westinghouse turbines and the MOD 5 were sold to private developers.

103. "Some Advice on Wind Farm Development," *EPRI Journal* (reprint of December 1992 issue):10–1.

104. Ibid.

105. Ros Davidson, "Limits Put On Hawaii Wind," *Windpower Monthly* 4 (September 1988):9.

106. Robert J. Hutchinson, "Transforming the Sea Wind," *Oceans* (September-October 1987):17.

107. Taylor Moore, "Excellent Forecast For Wind," *EPRI Journal* (June 1990):15 and trade literature from Zond Systems.

108. "The California Covenant," *Windpower Monthly* 5 (July 1989):18.

109. "No Quake Damage," *Windpower Monthly* 5 (November 1989):9.

110. "Wind Power Performance Reporting System," *1988 Annual Report*, CEC, P500-89-010, 1989); also see "California Commission Registers Great Improvement in Wind Statistics," *Windpower Monthly*, 5 (November 1989):9–10.

CHAPTER 12. A PERSPECTIVE ON THE FUTURE

1. Taylor Moore, "Forecast Excellent For Wind," *EPRI Journal* (June 1990):15. *EPRI Journal* is a publication of the Electric Power Research Institute, a think tank supported by companies that was established in 1972, devoted to improving electricity production, distribution, and consumption. In recent years it has supported wind energy.

2. Ros Davidson, "Public Perception Emerging as Core Issue for the Nineties," *Windpower Monthly* 5 (October 1989):12.

3. Christopher Flavin, "Windpower: A Turning Point," Worldwatch Paper 45, July 1981: 47.

4. See L. L. Freris, *Wind Energy Conversion Systems* (New York: Prentice Hall, 1990):389–90.

5. P. B. Bosley and K. W. Bosley, "Wind Generated Electricity: The Utility Industry's Perspective," in Windpower '91, *Proceedings*, American Wind Energy Association, Palm Springs, California, September 24–27, 1991: 390–7. Quote from page 390.

6. Ibid., 395.

7. Quoted in Devin Odell, "Wind Leads Renewables," *High Country News* 24 (June 29, 1992):15.

8. Information provided by Robert Lynette & Associates.

9. Jim Moriarty, CEO Carter Wind Turbine, Inc. to the author, February 1, 1993.

10. Thomas W. Lippman, "Future of Wind Power Gets a Lift," *Washington Post*, November 17, 1991, H-1, H-5. Also from U.S. Windpower trade literature.

11. Matthew L. Wald, "A New Era for Windmill Power," *New York Times*, September 8, 1992, C-1, C-6.

12. Ibid.

13. Harry Wasserman, "Nuclear Fade-Out," *The Nation* 258 (March 7, 1994):304.

14. "The Year 2000," brochure published by the American Wind Energy Association.

15. Interview with Paul Gipe, November 11, 1992.

16. David M. Eggleston and Forrest Stoddard, *Wind Turbine Engineering Design* (New York: Van Nostrand Reinhold, 1987:18.

17. Interview with Paul Gipe, November 11, 1992.

18. Ibid.

19. *EPRI Journal* (reprint of December 1992 issue): 13.

20. Cynthia Struzik, U.S. Fish and Wildlife Service, to Bruce Jensen, Alameda County Planning Department, February 5, 1993. The information was provided to the author by Darryl Mueller, of Save the Eagle. Struzik's letter was in response to U.S. Windpower's proposal for the removal, installation, and relocation of various wind turbines.

21. See Ros Davidson, "Research Incomplete on Lattice Towers and Birds," *Windpower Monthly* 10 (March 1994):19.

22. Trade literature on the Windjammer provided by Gordon Brittan, Montana Wind Turbine, Inc.

23. Telephone interview with Robb by the author, May 10, 1993.

24. Robert Thayer, Department of Environmental Design, University of California, Davis, to the author, February 25, 1992. Also see Thayer's *Gray World, Green Heart* (New York: John Wiley, 1994):275–6.

25. TV advertisements for the Lexus automobile and U.S. Air, both featuring wind turbines, are evidence of changing perceptions.

26. Brochure published by the Kern Wind Energy Assoc.

27. Donna Erickson, "Kids Are Blown Away by Wind and Its Power," *El Paso Times*, April 4, 1992, 2-E.

28. Wendell Berry, *The Gift of Good Land* (San Francisco: North Point, 1981):128.

29. U.S. Windpower trade literature.

30. These costs are summarized from Bernard Cohen's book, *It's Too Late: A Scientist's Case for Nuclear Energy* (New York: Plenum, 1983):233. Although Cohen was adamant in his support for nuclear power, he admitted that "wind can be made cost competitive, per average watt, with coal-fired or nuclear electricity": 245.

31. For instance see S.S. Penner and L. Icerman, *Energy* 1 (Reading, Mass.: Addison-Wesley, 1981):432–4. Penner and Icerman show the societal costs of coal mining to human life as well as the environment.

32. See scoping statement, Kenetech Windpower, Inc., Wind Energy Development Project, Carbon County, Wyoming. Cerca, January, 1994.

33. Peter Asmus, "New Technologies Revitalize Wind Power," *California Journal* (August 1992):413–415. Quote from page 414. Other articles on SMUD and wind energy by Asmus include "Electric Resource Planning: A Case Study," in John J. Kirlin, ed., *California Policy Choices* (Los Angeles: Univ. of Southern California, 1992):217–238; "Saving Energy Becomes Company Policy," *Amicus Journal* (winter 1993):38–42; and "Winds of Change," *Comstock's* (October 1992): 18–21.

34. Bill Dietrich, "Wind-power Farms Making Comeback in Pacific Northwest," *Seattle Times*, September 11, 1991, Sect. B; also see the *Oregonian*, November 16, 1991, and the *Seattle Times*, November 12, 1991, A-12.

35. Quoted in the *Fargo Forum* (North Dakota), region section, December 14, 1990.

36. Figures from Matthew L. Wald, "Putting Windmills Where It's Windy," *New York Times*, November 14, 1991, Sec. D.

37. *Horizons*, University of Texas at El Paso newsletter, April 17, 1992.

38. Conversation, Jeff Schroeter, Central and South West Corp., with the author, March 1994. Also see Ros Davidson, "Oil State Strikes Wind Potential," *Windpower Monthly* 10 (March 1994):20–1.

39. Ibid.

40. Steve Turek, manager, Wind Turbine Industries Corporation, to the author, February 4, 1993.

41. Karl H. Bergey, chairman, Bergey Windpower, to the author, January 18, 1993.

42. *Windletter* 6, June 1991: 3.

43. Derek Denniston, "Second Wind," *WorldWatch* 6 (March/April 1993): 33–35.

44. "1992 Wind Technology Status Report," a four-page flyer published by the American Wind Energy Association.

45. Denniston, "Second Wind," 35.

46. "1992 Wind Technology Status Report," AWEA.

47. Interview with Paul Gipe by the author, November 11, 1992.

48. Birger Madsen, "Where to Now?" *Windpower Monthly* 4 (November, 1988):6.

49. "PURPA Handbook for Independent Electric Power Producers," pamphlet published by the American Wind Energy Association, 1992: 9.

50. Interview with Robert Scheffler, SCE, by the author, November 18, 1992.

51. Interview with Paul Gipe by the author, November 11, 1992; also interview with Curt Maloy, November 9, 1992.

52. Gerald W. Braun and Don R. Smith, "Commercial Wind Power: Recent Experience in the United States," *Annual Review of Energy* 17 (1992):113

53. O. Hohmeyer, *Social Costs of Energy Consumption* (Berlin: Springer-Verlag, 1988). As cited in Braun and Smith, "Commercial Wind Power," 13.

54. For much greater detail on this question, see Pace University's *Environmental Costs of Electricity* (New York: Oceana, 1991).

55. "Comments of the American Wind Energy Association," submitted by Michael Marvin to the Texas Public Utilities Commission, January 15, 1993. Transcript copy provided by the American Wind Energy Association, Washington, D.C.

56. Madsen, "Where to Now?" 6.

57. Statistics from "Comments," January 15, 1993.

EPILOGUE

1. Paul Gipe, *Wind Energy Comes of Age* (New York: John Wiley, 1995), figures from page 72.

2. Ibid.:v.

3. See Peter Passell, "A Makeover for Electric Utilities," *New York Times*, February 3, 1995, C1; also Agis Salpukas, "70s Dreams, 90s Realities," *New York Times*, April 11, 1995, C1.

4. The activities of the California Public Utility Commission, as described in chapter 9, provide an example of an expanded environmental mission.

BIBLIOGRAPHICAL NOTE

THIS BRIEF NOTE IS LIMITED TO GENERAL SOURCES ON THE SUBJECT OF wind energy. It is intended to point the researcher toward avenues for exploration, not to specific citations. It is not comprehensive, but indicates sources that I found useful in preparing this work. Readers interested in specific references can refer to the chapter notes. The notes include suggestions for further reading and research into specific subjects.

The history of wind energy may reasonably be broken into two chronological periods. The first, from about 1880 to 1970, lends itself to a focus on individual wind plants, largely in rural settings. The second, from 1970 to the early 1990s, is contemporary, not only as to the time period but also in dealing with the function of producing bulk energy for electric grid systems. This brief note discusses some of the resources available for these two periods.

THE EARLY PERIOD, 1880 TO 1970

No bibliographies are available for the early period of wind energy. T. Lindsay Baker's *A Field Guide to American Windmills* (1985) contains a massive bibliography of sources on water-pumping windmills, but is limited in its citations for electric wind-turbines. Although there is a bibliography on wind energy, it considers the pre-1970 period to be practically nonexistent. We are therefore left with, as the best sources for the early part of the twentieth century, the *Agricultural Index* and the *Engineering Index*. These multivolume indexes stretch back to the turn of the century, and are arranged so that wind-energy articles are listed separately and given a brief abstract. The abstract is particularly useful in determining whether to pursue any given article. The late nineteenth-century issues of the *Scientific American* also provide a lot of material, and in useful form. The *Scientific American* acted as a clearinghouse for inventors and it is indexed for each six-month period.

Two specialized libraries are particularly helpful on the topic. Perhaps the most worthwhile is the Engineering Library in New York City, located across the street from the United Nations building. The library, funded by U.S. engineering societies, holds a complete run of many esoteric engineering or engineering-related magazines and journals, domestic and foreign. For the agricultural aspect of wind energy, the National Agricultural Library in Maryland is unsurpassed. This multistoried library, administered by the Department of Agriculture, holds many obscure journals.

Among university libraries with significant collections, the Freiberger Library, Case Western Reserve University, Cleveland, is noteworthy for the C. R. Brush Collection, invaluable in understanding the scientist and inventer Charles Brush and the building of the Brush wind dynamo, the world's first wind-electric converter. For primary material on small-farm wind-energy systems between 1920 and 1950, the Windmill Literature Collection of the Panhandle-Plains Historical Museum, Canyon, Texas must be consulted. Both the Southwest Collection at Texas Tech University and the Huntington Library at San Marino, California, provide materials unavailable elsewhere. Few wind-energy company papers have survived. However, the Colorado Historical Society, in Denver, has catalogued the Fritchle Wind-power Electric Manufacturing Company papers. This record of Oliver Fritchle's disheartening effort to establish a wind-energy company is instructive, particularly regarding the early history of the industry.

THE CONTEMPORARY PERIOD: 1970 TO THE 1990s

The resurgence of interest in wind energy in the early 1970s produced a helpful bibliography. Barbara L. Burke and Robert N. Meroney, *Energy From the Wind: Annotated Bibliography* (Fort Collins, CO: Solar Energy Application Laboratory, 1975, 1979, 1982) is a basic bibliography with three supplements.This work is an excellent start, but it is far from complete and for the historic period is simply ineffectual. SERI (Solar Energy Research Institute) has produced basic bibliographies as a part of its mission to promote alternative energy. But a basic bibliography for the crucial 1980s decade has not, it seems, yet been compiled.

The American Wind Energy Association (the trade organization for wind-energy manufacturers, developers, and operators) has produced significant materials. These can provide leads to useful contemporary publications. The association periodically provides members with nationwide newspaper clippings on wind energy and the monthly *Windletter*, giving industry news.

The most useful library on wind energy—funded by SERI, now reconstituted as NREL (National Renewable Energy Laboratory)—is in Golden, Colorado. The library makes no pretense to hold historic materials, but it does catalog most contemporary items. It is open to the public. T. Lindsay Baker (Baylor University) has a personal collection and files that are extensive for the historic period, and Don Smith, a wind-energy consultant for the Pacific Gas and Electric Company, has a Berkeley flat crammed with contemporary materials. Whatever has been published in the last twenty years will likely be found in Smith's cache.

Trade literature is particularly helpful in chronicling recent events. The following consultants, companies, and associations have useful materials: U.S. Windpower, Inc. (Kenetech), Zond Systems, Inc., WindMaster, Incorporated, Montana Wind Turbine, Inc., Wind Turbine Industries, Inc., Minnesota Windpower Inc., Bergey Windpower Co., International Wind Systems, Inc. (Carter), Wind Baron Corporation, Wind Harvest Corporation, World Power Technologies, Inc., Pacific Gas and Electric Co., Southern California Edison Co., R. Lynette & Associates, Robert D. Kahn Associates, the Kern Wind Energy Association, the Desert Wind Energy Association, and the American Wind Energy Association.

As with the earlier period, so it is with gaining an understanding of the contemporary situation in California and elsewhere: oral interviews are indispensable. People listed at the front of this book in my acknowledgements of help received are valuable sources. I also again want to mention that often neglected source—clippings files on wind energy. There are collections of these at the libraries of Palm Springs, California, and Bakersfield, California. At Livermore, California, an index of local newspapers on many subjects, including wind energy, makes it possible to research both news and public opinion on the subject.

INDEX